GRAVITY'S FATAL ATTRACTION

Black Holes in the Universe

Second edition

Mitchell Begelman
University of Colorado, Boulder

Martin Rees
University of Cambridge

CAMBRIDGE
UNIVERSITY PRESS

CAMBRIDGE UNIVERSITY PRESS

Cambridge, New York, Melbourne, Madrid, Cape Town, Singapore, São Paulo, Delhi

Cambridge University Press

The Edinburgh Building, Cambridge CB2 8RU, UK

Published in the United States of America by Cambridge University Press, New York

www.cambridge.org

Information on this title: www.cambridge.org/9780521717939

First published 1995
Second edition published 2010

Printed in the United States

A catalogue record for this publication is available from the British Library

ISBN   978-0-521-88944-5 Hardback
ISBN   978-0-521-71793-9 Paperback

GRAVITY'S FATAL ATTRACTION
Black Holes in the Universe
Second edition

Richly illustrated with the images from observatories on the ground and in space, and computer simulations, this book shows how black holes were discovered, and discusses our current understanding of their role in cosmic evolution.

This second edition covers new discoveries made in the past decade, including definitive proof of a black hole at the center of the Milky Way, evidence that the expansion of the Universe is accelerating, and the new appreciation of the connection between black holes and galaxy formation. There are entirely new chapters on gamma-ray bursts and cosmic feedback.

Begelman and Rees blend theoretical arguments with observational results to demonstrate how both have contributed to the subject. Clear, explanatory illustrations and photographs reveal the strange and amazing workings of our Universe. The engaging style makes this book suitable for introductory undergraduate courses, amateur astronomers, and all readers interested in astronomy and physics.

Mitchell Begelman is Chairman of the Department of Astrophysical and Planetary Sciences and Fellow of JILA, at the University of Colorado at Boulder. He has won several awards, including the Guggenheim Fellowship, Sloan Research Fellowship, and the American Astronomical Society Warner Prize.

Martin Rees is Professor of Cosmology and Astrophysics and Master of Trinity College at Cambridge University, and Astronomer Royal. He is a winner of the Gold Medal of the Royal Astronomical Society and other awards. He has written several books for a general readership.

The authors are leading authorities on the astrophysics of black holes, and were key players in establishing the existence of black holes and their role in cosmic phenomena.

Contents

Preface to the Second Edition

As early as 1967 black holes were creeping into public consciousness. On the television program *Star Trek*, Captain Kirk and the crew of the starship *Enterprise* were caught in the gravitational field of an "uncharted black star" and were propelled backward in time. Astronomers then had no clue whether black holes were real or just theoreticians' constructs. Certainly no serious evidence of their reality existed, and few would have guessed that they would soon become the object of intense astronomical study. Until the 1970s, the black hole was still a novel concept, studied by specialists in Einstein's general theory of relativity. And that theory itself (though already several decades old) had only tentative empirical support. Gravity, one of the fundamental forces of nature, was still poorly understood.

Thanks to a technological revolution in observational astronomy – new detectors and mirror designs in optical telescopes, radio telescopes that offer far sharper images than even the best optical instruments, and observations made from space, revealing the sky at infrared, ultraviolet, X-ray, and gamma-ray wavelengths – we are now confident that there are millions of black holes in every galaxy. Each of these holes is the remnant of an ordinary star several times more massive than the Sun. More remarkably, giant black holes, weighing as much as millions (or even billions) of suns, lurk in the centers of most galaxies. The most energetic phenomena in the Universe – quasars, and jets a million light-years long erupting from the centers of galaxies – are powered by black holes. The same phenomena, in miniature, are energized by the smaller holes within our own Galaxy.

This book describes the extraordinary ways in which black holes make their presence known. We discuss the designs and accidents through which they were discovered, and how far we have come toward understanding their relationship to other structures in the cosmos. Every advance in technology has disclosed an assortment of dazzling, unexpected phenomena. Some of these phenomena, like "gravitational lenses," are well understood and are being co-opted as tools in the search for black holes and other forms of "dark matter"; others, such as gamma-ray bursts, remain poorly understood. In between lies a range of phenomena for which understanding is substantial but incomplete: the processes and flow patterns around black holes; how the first ones formed; and the two-way interactions between supermassive holes and their "host" galaxies. And new questions for the future are constantly arising. Can gravitational waves from merging black holes be detected? How much could black holes contribute to the "dark matter" in the Universe? Could microscopic holes exist – the size of an atomic nucleus but the weight of a mountain? As the Universe evolves, what is the ultimate fate of matter "swallowed" by black holes?

As theorists in a profession heavily dominated (and rightly so) by observers, we demand empirical proof that black holes exist, and that they have the properties predicted by Einstein's theory. Is there a "smoking gun" that removes all doubt of their existence, or of their ubiquity at the level we claim? When the first edition of the book appeared (in 1995) there were still some doubts. But the evidence has strengthened tremendously since then, and our understanding of the role of black holes in the Universe has deepened as well. In this completely new edition we have substantially rewritten several chapters and included much new material. There are new chapters on gamma-ray bursts and cosmic feedback. Images, from observations and simulations alike, have been updated throughout. These advances are owed to new and more sensitive instruments and to vastly more powerful computers than were available in the 1990s.

Black holes are an important feature of our cosmic environment. We are beginning to understand the exotic ways they manifest themselves, and how astronomical observations can probe their immediate surroundings.

These cosmic "fireworks," spectacular as they are, may ultimately prove to be of greatest value as stepping-stones to even more profound knowledge. Shrouded from view, deep inside black holes, lurk mysteries that will not be understood without a unification of Einstein's theory of gravity with the other great "pillar" of twentieth-century physics, the quantum theory. Such new insights, when they are achieved, may change our view of the nature of time, and of our entire Universe. The strong new evidence that black holes indeed exist strengthens the motivation for this fundamental quest.

Research in astrophysics is, for us, an intensely interactive activity. Many observers have generously discussed their latest work with us; we have also benefited from other theorists who have special knowledge and expertise that we lack. We especially acknowledge all we have learned from Roger Blandford, a frequent collaborator throughout the last thirty years. We are fortunate to have, among our immediate colleagues, leading authorities such as Andrew Fabian, Stephen Hawking, Donald Lynden-Bell and Richard McCray. We are grateful to them, and to our many other collaborators. We are also grateful to many colleagues who took the time to prepare illustrations – often of results "hot off the press," or even prepublication.

We have been especially fortunate in our publisher and editors – Jonathan Cobb, Susan Moran and Nancy Brooks of Freeman helped with the first edition. And we remain grateful to Colin Norman, Joseph Silk, and Virginia Trimble for offering helpful comments on that edition.

Preface to the Second Edition

We are grateful to our respective staffs – especially Judith Moss in Cambridge and Elaine Verdill and her colleagues in Boulder – for keeping the three-way exchange of text and pictures running smoothly during those (more or less) preinternet days. During the preparation of this new edition, we have been helped and encouraged by Simon Mitton and Vince Higgs at Cambridge University Press. We are also grateful to Lindsay Barnes, Laura Clark and Chris Miller for editorial assistance with both text and pictures; and to Phil Armitage, Davide Lazzati and Chris Reynolds for valuable advice on the updates.

1 GRAVITY TRIUMPHANT

Gravity is the one truly universal force. No substance, no kind of particle, not even light itself, is free of its grasp. It was Isaac Newton who, over three hundred years ago, realized that the force holding us to the ground, and governing a cannon ball's trajectory, is the same force that holds the Moon in its orbit around the Earth. This force, which later came to be called gravity, is the mutual attraction among all bodies. Newton showed how the motion of each planet in the Solar System is the combined outcome of the gravitational pull of the Sun and all the other planets, each contributing according to its mass and distance from the others. Out of Newton's conceptually simple prescriptions came the calculations that have guided spacecraft to all the planets, told us a year in advance that Comet Shoemaker–Levy would hit Jupiter, and enabled us to determine the mass of the Milky Way Galaxy. On cosmic scales gravity dominates over every other kind of force. Every significant level of structure in the Universe – stars, clusters of stars, galaxies, and clusters of galaxies – is maintained by the force of gravity.

Nowhere is gravity stronger than near the objects we call black holes. The effect of gravity at the surface of a body intensifies if the mass of the body is increased, or its size is decreased. Gravity's strength can be characterized by how fast a rocket must be fired to escape from the body. For the Earth, the escape speed is 40000 kilometers per hour; for the Sun it is nearly a hundred times greater. This tendency for the escape speed to become ever larger for more massive, or more compact, bodies raises an obvious conceptual problem. The gravitational pull of the Sun dominates the Solar System, yet its escape speed is only 1/500 the highest speed possible, the speed of light. What would be the effect of gravity around a body for which the escape speed was as high as that of light?

The Reverend John Michell, an underappreciated polymath of eighteenth-century science, puzzled over this question in 1784. Arguing on the basis of Newton's laws and assuming that light behaves like particles of matter (the view favored by Newton and his contemporaries), Michell wrote a paper in which he noted, "If the semi-diameter of a sphere of the same density as the Sun were to exceed that of the Sun in the proportion of five hundred to one, and supposing light to be attracted by the same force in proportion to its *vis inertiae* [inertial mass] with other bodies, all light emitted from such a body would be made to return towards it, by its own proper gravity." With remarkable foresight, Michell was suggesting, as did the French astronomer and mathematician Pierre-Simon Laplace a decade later, that the most massive objects in the Universe might be undetectable by their direct radiation but still manifest gravitational effects on material near them. Nearly two hundred years later, astronomers discovered objects for which gravity is as strong as Michell envisaged – these powerfully gravitating bodies are neutron stars and black holes.

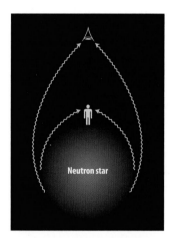

The severe bending of light by gravity means that an observer on the surface of a neutron star can see farther to the horizon. An observer high above the star could see much more than half the surface.

Neutron stars, discovered in 1968, are objects slightly heavier than the Sun but extending only 10 to 20 kilometers across, with an escape speed about half that of light. While light can escape from a neutron star, its path is strongly curved by gravity, just like the trajectory of a rocket boosting a spacecraft into interplanetary space from the Earth. The extreme strength of gravity around a neutron star leads to consequences far removed from common-sense experience. Because light itself follows a strongly curved trajectory, an observer's perspective would be severely distorted near a neutron star. One would be able to see much farther to the horizon, and from high above the surface, more than half the entire star would be visible. So powerful is gravity that a pen dropped from a height of one meter above a neutron star would release as much energy as a ton of dynamite.

Conditions near a neutron star are not quite as extreme as Michell described for his hypothetical light-retaining body. But if a neutron star were a few times smaller, or just slightly heavier, it would trap all the light in its vicinity. It would then be a black hole. Black holes are not merely stars whose surface is hidden from view by the bending of light. In 1905, Einstein's special theory of relativity established that the speed of light sets a limit to the speed at which any kind of matter or signal propagates, so that the kind of object envisaged by Michell would be completely cut off from the outside Universe. Nothing that ventures inside it would ever be able to escape. As it turns out, in a black hole gravity so overwhelms other forces that matter is crushed virtually to a point. These bizarre and counterintuitive properties led many researchers to suppose that, somehow, matter could never become quite so condensed that it produced a black hole. But we now know that black holes are not mere theoretical constructs; they exist in profusion and account for many of the most spectacular astronomical discoveries of recent times.

This book describes how black holes were found and what their existence implies for cosmic evolution and our understanding of gravity. The first candidate black holes were discovered soon after neutron stars. Since then, evidence has mounted that millions of black holes exist in our Milky Way Galaxy, each with a mass roughly 10 times that of the Sun. And in the centers of most galaxies, there almost certainly exist monster black holes weighing as much as 100 million suns, similar in mass and radius to the objects conjectured by Michell.

General Relativity and Black Holes

Neither Newtonian gravity nor Newton's laws of motion are applicable to situations where the gravitational fields are as strong as those conjectured by Michell and Laplace. On the Earth, as elsewhere in the Solar System,

Newton's theory of gravitation affords an excellent approximation. But where gravity is so strong that it can accelerate matter to nearly the speed of light – approximately 300 000 kilometers per second – the Newtonian description no longer works. We cannot formulate a consistent picture of a black hole without deeper insights into gravity than Newton's theory offers. Fortunately, long before black holes were discovered, Einstein's general theory of relativity had provided such insights and laid a firm theoretical foundation for our understanding of black holes. Einstein's theory has been verified by increasingly precise experimental tests, and its equations have solutions that represent situations where gravity is overwhelmingly strong, as it is in black holes.

The papers that established Albert Einstein as the greatest physicist since Newton are all contained in just two volumes of the journal *Annalen der Physik,* published in the years 1905 and 1916. The old backnumbers of scientific journals, relegated to remote stockrooms in academic libraries, are usually seldom consulted, but these two particular volumes are rare and valued collector's items. In 1905, the 26-year-old Einstein enunciated his "special theory of relativity," which (among other breakthroughs) emphasized the special significance of the speed of light and postulated the fundamental interconvertibility or "equivalence" between mass and energy popularly paraphrased in the expression $E = mc^2$. He also proposed that light is quantized into "packets" of energy and formulated the statistical theory of Brownian motion, the frenetic dance of dust particles suspended in a liquid or gas. These contributions alone would cause Einstein to be ranked among the half-dozen great pioneers of twentieth-century physics.

But it is his gravitational theory, "general" relativity, developed eleven years later, that puts Einstein in a class by himself. Had he written none of his 1905 papers, it would not have been long before the same concepts would have been put forward by some of his distinguished contemporaries: the ideas were "in the air." Well-known inconsistencies in earlier theories and puzzling experimental results would, in any case, have focused interest on these problems. General relativity, however, was not a response to any particular observational enigma. Motivated by more than the desire to explain concrete observations, Einstein sought simplicity and unity. Whereas Newton conceived gravity as a force instantaneously transmitted between bodies – a view clearly inconsistent with the speed limit on the propagation of signals – Einstein postulated that gravitational fields are manifestations of the curvature of space itself. Masses do not "exert" a gravitational pull, deflecting bodies from a straight path. Rather, their presence distorts the space within and around them. According to general relativity, bodies moving through space follow the straightest path possible through an amalgam of space and time called "spacetime." But when space is distorted, these paths become curved and accelerating

Albert Einstein in 1905, aged 26. In that year he proposed the special theory of relativity, deduced the quantum nature of light, and explained Brownian motion of dust particles.

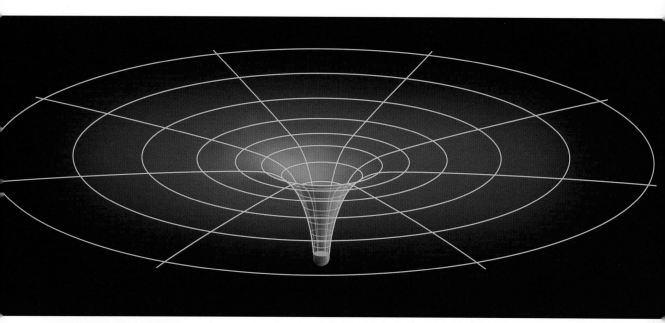

According to the general theory of relativity, a gravitating body stretches and distorts the space around it, much like a lead weight resting on a rubber membrane. Particles following the straightest possible paths in curved spacetime appear to be responding to an attractive force. If the gravitating object represented were a black hole, the depression would be infinitely deep.

trajectories that we might interpret as the reaction to a force. In the words of the noted relativist John Archibald Wheeler, "Space tells matter how to move; matter tells space how to curve."

As Einstein himself said when he announced his new work, "Scarcely anyone who has fully understood the theory can escape from its magic." The physicist Hermann Weyl described it as "the greatest example of the power of speculative thought," and Max Born called it "the greatest feat of human thinking about Nature." Had it not been for Einstein, an equally comprehensive theory of gravity might not have arrived until decades later, and might well have been approached by quite a different route. Einstein's creativity thus put a uniquely individual and long-lasting imprint on modern physics. Most scientists hope that their work will last, but Einstein is unique in the degree to which his work retains its individual identity.

Indeed, general relativity was proposed so far in advance of any real application that it remained, for forty years after its discovery, an austere intellectual monument – a somewhat sterile topic isolated from the mainstream of physics and astronomy – whose practitioners, according to Thomas Gold, were "magnificent cultural ornaments." This remoteness from the mainstream stands in glaring contrast to its more recent status as one of the liveliest frontiers of fundamental research. Its reputation for being an exceptionally hard subject was sometimes exaggerated, not least by its early practitioners. (When Arthur Eddington was asked if it were true that just three people in the world understood general relativity, he

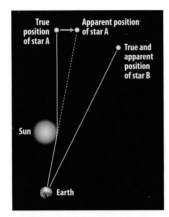

During the 1919 solar eclipse, Arthur Eddington observed the deflection of starlight by the Sun's gravity. The effect is tiny; according to Einstein's theory the maximum angle of deflection would be 1.75 seconds of arc (1/2000 of a degree) for a light ray grazing the Sun's surface.

is rumored to have wondered who the third might be.) The theory is now taught almost routinely to physics students, along with quantum physics, electromagnetism, and the rest of the physics canon.

To be sure, general relativity also had its share of early experimental successes. It accounted for long-recognized anomalies in Mercury's orbit, and was famously confirmed by the results of the 1919 solar eclipse expedition led by the British astrophysicist Arthur Eddington. Eddington's observations revealed that light rays passing close to the Sun were deflected by just the amount predicted by Einstein's theory. (A Newtonian argument, as might have been used by Michell, would have predicted half the observed amount.) For slowly moving bodies in weak gravitational fields, the predictions of general relativity agree almost exactly with those of Newton's laws – as they must, given that the latter have withstood three hundred years of testing. In the Solar System, the predicted departures from Newtonian theory amount to only about one part in a million. But experiments, including the tracking of space probes and higher-precision versions of Eddington's light-deflection measurements, are now precise enough to determine these tiny differences within a better than 1 percent margin of error. General relativity has been verified in every single case and, moreover, has proved more accurate than other, similar theories.

The recent revival of interest in gravitational physics has been inspired not only by the high-precision measurements whose results have favored Einstein's theory over its rivals, but also – to a greater extent – by dramatic advances in astronomy and space research. In neutron stars and black holes, the distinctive features of general relativity take center stage; they are not merely trifling modifications to Newtonian theory. These objects offer us the opportunity to test Einstein's theories in new ways. The other arena where relativity is crucial is cosmology – the study of the entire expanding Universe. During the past forty years, cosmology has evolved from a subject about which we knew little more than Hubble's law of universal expansion (which states that distant galaxies appear to recede from us at a speed that is proportional to their distance) to a science that allows us to infer cosmic history reliably back to one second after the big bang, and to make compelling conjectures about still earlier eras. The entire fate of the Universe depends on the relativistic curvature of space induced by everything in it. And black holes may have played a special role in the evolution of the Universe, as we shall show in late chapters.

From a relativistic point of view, the strength of a gravitational field can be characterized by the ratio of the velocity required to escape from the field (the "escape speed" often discussed in the context of space exploration) to the speed of light, all squared. When this ratio is very small, as it is near the Sun (1/250 000), Newtonian theory provides a near-perfect description of gravity. But when this ratio approaches 1, as it does close to a neutron star or black hole, general relativity predicts very bizarre and dramatic effects. We have already mentioned the bending of light rays, which becomes extreme as they traverse the curved space associated with a strong gravitational field. Perhaps even stranger is the "time dilation" effect – the apparent slowing down of time in the presence of gravity. If space were marked out by a set of fixed clocks, those located where gravity is strong would seem to a distant observer (located where gravity is weak) to run slow; conversely, distant clocks would seem to run fast when viewed by observers near the central mass. No matter how much the rate at which time passes varies from place to place, at a given spot time would pass at the same rate for *every* process or event. If we could send astronauts to the surface of a neutron star, and listen to them through a radio or view them through a telescope, their speech and movements would appear to us to be about 25 percent slower than normal. As far as they were concerned, however, they would be operating at normal speed and our messages would seem speeded up. Any oscillation of the kind that generates an electromagnetic wave would behave just like the ticks of a clock. To a distant observer watching such an oscillation near the surface of a large mass, the interval from one oscillation to the next would appear longer, and so would the wavelength of the corresponding electromagnetic wave. Because in the optical spectrum red light is longer in wavelength than the other colors, the lengthening of wavelength is called a redshift. Thus light emitted close to the central mass would display a "gravitational redshift" when observed farther out; in other words, its observed wavelength is longer than the wavelength when it was emitted.

Black holes exhibit the effects of "strong" gravity in its purest, and most extreme, form. The possible existence of objects that have collapsed to such small dimensions that neither light nor any other signal can escape from them is predicted by most theories of gravity, not merely by Einstein's general relativity. But it is within the context of general relativity that the most complete understanding of black holes and their properties has been developed.

The first theoretical description of a black hole within the framework of general relativity was given in 1916 by Karl Schwarzschild, who calculated the distortions of space outside a spherically symmetrical body of a given mass. An observer, sitting at a fixed distance from the center of the sphere, would measure the same curvature of space (and therefore the same gravitational strength) regardless of the body's radius. However, if the body were smaller than a certain size, it would not be seen, even though its gravity would still be felt. This boundary between visibility and invisibility, or "horizon," corresponds to the minimum radius from which light or anything else can escape to an external observer, or, almost equivalently, the radius at which

Karl Schwarzschild in his academic robe of the University of Göttingen, Germany. By the winter of 1915–16, he had voluntarily taken leave from his position as director of Potsdam Astrophysical Observatory for active duty in World War I. He was dying of a rare disease contracted on the eastern front when he discovered the Schwarzschild radius.

the gravitational redshift is infinite and the escape speed is equal to the speed of light. Objects (and light) can cross the horizon in the inward direction, but nothing can come out. Any spherical body smaller than this radius is a "Schwarzschild black hole." (Schwarzschild's equations are discussed in the box below.)

The horizon of a Schwarzschild black hole lies at a radius of $3M$ kilometers, where M is the mass of the black hole measured in solar masses (units of mass each equivalent to the mass of our Sun). Coincidentally, this is exactly the same radius that Michell and Laplace would have calculated using the incorrect Newtonian theory.

The Schwarzschild Metric and the Curvature of Space

According to Newton, the gravitational force outside a spherical object is inversely proportional to the square of the distance from the object's center. From this finding, it follows that orbits around such an object must be ellipses. In 1916, Karl Schwarzschild determined how gravity would behave outside a spherical object according to Einstein's theory. The results are expressed as the *Schwarzschild metric*, describing the curvature of space outside a spherically symmetric mass.

Where gravity is weak and the orbital motion is very much slower than the speed of light, the predictions of Schwarzschild's metric agree almost exactly with Newton's. The Earth moves around the Sun at about one ten-thousandth the speed of light, and Newton's theory is an excellent approximation for all orbits in the Solar System. According to general relativity, however, orbits around a central mass don't close up as exact ellipses, as Newton's theory predicts. Whereas Newtonian theory predicts that the orientation of an elliptical orbit remains fixed in space, Einstein's theory predicts that the direction of the "long axis" of the ellipse gradually rotates. The expected

effect has been detected in Mercury's orbit (though the "Newtonian" perturbations due to the other planets are much stronger). Another consequence of the Schwarzschild metric is that light moving through the Solar System is deflected twice as much as it would be if you simply applied Newton's laws to a "particle" moving at the speed of light. This effect, though still very small, has been measured accurately.

Einstein's theory has remarkable consequences when gravity is strong – wherever, in Newton's theory, the orbital speeds would be a good fraction of the speed of light. Here the underlying concept that particles take the "straightest" path through curved space – which offers a deeper insight into gravity even when it is weak – becomes indispensable.

To understand the concept of "curvature," it is helpful to think first of how we could check whether an ordinary two-dimensional surface was curved or flat. Suppose you drew two circles of latitude around the North Pole of the Earth, one slightly larger than the other, and measured their circumferences by crawling round them. If the Earth were flat,

the difference would be $2\pi x$, when x is the difference between the radii. However, on the Earth, where x is measured along a meridian of longitude, the difference would be less than $2\pi x$. Measurements of this kind can tell two-dimensional beings whether the surface they live on is curved.

The curvature of three-dimensional space around a point mass can be understood in the same way: the circumferences of circles

Circle   **Sphere**

Concentric circles drawn on a plane differ in circumference by an amount equal to 2π times their difference in radius. If we drew concentric circles on a sphere and measured their radii along the surface of the sphere, the difference in circumference would be smaller than 2π times their difference in radius. This discrepancy arises because the sphere is curved, whereas the plane is flat.

Whereas the gravitational deflection of light is a very small effect in the Solar System and in most astronomical contexts, light is severely bent in the strongly curved space close to the Schwarzschild radius. Indeed, light is even more strongly perturbed than around a neutron star. If our astronauts could hover just outside the Schwarzschild radius, they would need to aim their radio beam almost directly outward to keep it from being dragged back and eventually swallowed by the black hole. If they ventured within this radius, they would be unable to send any signals whatever to the outside world.

aren't $2\pi r$, where r is the radial distance. Schwarzschild's equations tell us what the curvature is – how the circumference differs from $2\pi r$, and how it depends on radius.

According to these equations, a bizarre thing happens to the circumference when the radius reaches $3M$ kilometers, where M is the central mass in solar masses. Instead of continuing to decrease with decreasing radius, as we would expect,

the circumference levels off and becomes independent of radius! This means that, to a distant observer, an object can keep falling and falling toward the center of gravity and never appear to get any closer. In other words, its inward motion seems to "freeze out" when it reaches $3M$ kilometers. The reader has probably noted that this radius is precisely the radius of the black hole's horizon; indeed, what we have just described is an interpretation of

the horizon in terms of the curvature of space.

Where does the numerical value of the horizon's radius come from? Using Schwarzschild's equations, a physicist might write the "Schwarzschild radius" in the form $2GM/c^2$. Here, M is the central mass, c is the speed of light, and G is the "universal constant of gravitation" from Newton's theory.

Schwarzschild's equations imply that gravity, in effect, increases with decreasing distance more steeply than the inverse square law of Newton – it becomes more than 4 times stronger when the distance halves. A particle close to the central mass needs more energy than in Newtonian theory in order to stay in orbit. When an object is orbiting at less than 3 times the Schwarzschild radius, this effect is so drastic that a slight nudge to a body would send it spiraling toward the center. (In Newtonian theory, conservation of angular momentum would prevent this from happening.) Indeed, at 1.5 Schwarzschild radii, a body would have to move at the speed of light to stay in orbit.

Close to a Schwarzschild black hole, space is severely stretched in the radial direction compared to the perpendicular directions. Because of the curvature of space, an object falling inward never seems to get any closer to the center of the hole than the radius of the horizon. This three-dimensional effect can be represented in two dimensions by a so-called embedding diagram that resembles the mouth of a trumpet.

Trajectories of light rays become severely curved close to the horizon of a black hole. As the horizon is approached, a light beam must be aimed within an increasingly narrow cone in order to avoid being dragged into the hole. At 1.5 Schwarzschild radii a carefully aimed light beam can orbit the black hole indefinitely. At the horizon, the escape cone closes up – no light rays emitted from this radius can avoid being sucked into the black hole. Radiation from just outside the Schwarzschild radius would reach a distant observer with a large redshift, and a clock near the hole would appear to run slow.

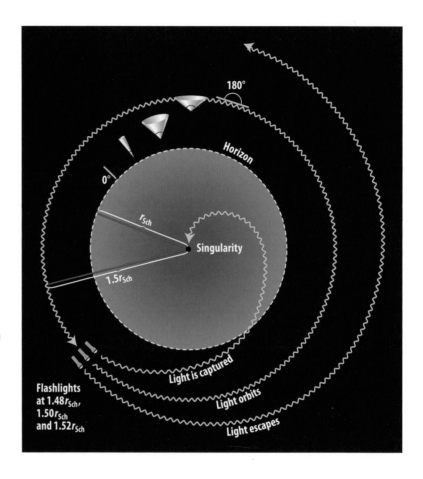

Although the region enclosed within the horizon is shrouded from an external observer's view, the explorers would initially find nothing unusual as they passed through the critical radius. However, they would then have entered a region from which, however hard they fired their rockets, they could never escape. They would be inexorably drawn to a central point – the "singularity" – where tidal forces (the difference in the gravitational acceleration from one side of their spaceship to the other) would become infinite, and they and their ship would be crushed out of existence. As measured by their own clock, the trip from the horizon to the singularity would have taken just about as long as the time required for a light beam to travel the length of the Schwarzschild radius. But an external observer would not even see the astronauts cross the Schwarzschild radius: as they approached the horizon, their clock would appear to the distant observer to run slower and slower, and any signal that they sent would become more and more redshifted. Photons –

packets of electromagnetic energy carrying the signal – would arrive less and less frequently, even as they weakened individually, until the signal practically ceased. Soon it would become undetectable, even by the most sensitive of receivers. As the redshift continues growing, it doubles in the length of time light takes to travel along the Schwarzschild radius. For a black hole with the mass of the Sun, this time is so short that the astronauts would disappear and their signals fade in less than a millisecond. Thus, one cannot from a safe distance observe anything about the extreme conditions near the central singularity. To learn about the interior of a black hole, one must feel a Faustian urge sufficiently strong to make one venture inside the horizon despite the inevitable destruction this implies!

Schwarzschild's theory of the horizon was at first regarded as a mathematical curiosity, but surely not a representation of a real object; then, in 1939, J. Robert Oppenheimer and Hartland Snyder suggested that the implosion of a dying star might actually create a black hole. Their conjecture generated surprisingly little interest. Physicists were critical of the highly idealized nature of the calculation through which Oppenheimer and Snyder showed what would happen to a collapsing star. In order to handle the mathematics without access to electronic computers (which had not yet been invented), they were forced to assume that the collapsing star was perfectly spherical and nonrotating, without the slightest blemish or imperfection. This, of course, was an outrageously unrealistic assumption. Most physicists who reflected on the result had the "gut feeling" that any slight deviation from perfect symmetry would be grossly amplified during the collapse. Different parts of the infalling star would "miss" one another, no horizon would form, and the result would be some indeterminate mess but certainly not a black hole. Another reason for the early dismissal of black holes was the troubling presence of a singularity, a place where (according to the equations) some physical quantity becomes infinite. Nature is simply not supposed to behave this way, so a singularity is usually interpreted as evidence that physicists, not nature, made a mistake!

We now understand that the equations of general relativity do indeed predict that black holes must contain a singular point (inside the horizon, and thus hidden from our view) where gravitation is infinitely strong. And physicists generally agree that general relativity's prediction of a singularity signals the theory's inability to cope with situations where gravitational effects vary over too small a distance. It is hoped these problems will be remedied in the future, once the theory of gravity is combined successfully with quantum theory, which treats phenomena occurring on tiny scales. But it is not thought that the modifications of such a "unified theory" will have much impact on our understanding of the black holes discovered in nature. Indeed, the real black hole singularity – the one at the center – was *not* the one that disturbed physicists between 1940 and 1960. They were worried about what *seemed* to be a singularity at the radius of the horizon, where time apparently came to a standstill. We now know that the latter is a singularity in Schwarzschild's *description* of the physics, and not in the physics itself.

Addressing the problem of asymmetry in the collapse proved to be more subtle. Its final resolution, in the late 1960s and early 1970s, was the key piece of theoretical progress in general relativity that finally convinced many astrophysicists that black holes were objects that might actually exist. When an exactly spherical body collapses, even Newtonian theory predicts that everything crashes together at the center – but nobody worried about singularities in Newtonian theory, because any deviation from symmetry would prevent the collapse focusing onto a single point. However, Stephen Hawking and Roger Penrose showed, in the late 1960s, that Einstein's theory always leads to a singularity surrounded by a horizon, even when the black hole forms as the consequence of a realistic and irregular collapse. Moreover, the resulting black hole, as viewed from outside, has remarkably simple and standardized properties. As a black hole forms, any residual asymmetry in its structure is "radiated away" in a burst of gravitational waves (explained in Chapter 10), and the black hole quickly settles down to a stationary state. In this state, the hole reveals itself to us only by its external gravitational field, which is characterized by only two quantities, the hole's mass and a quantity called angular momentum that is related to the black hole's rate of spin. The reduction of a black hole to two simple properties is colloquially described as the theorem that "black holes have no hair." (In principle, a black hole could also have a net electric charge, but it is unlikely that any realistic collapse could produce a black hole with significant charge because that charge would be neutralized by attracting an equal and opposite charge from its surroundings.)

One more notable piece had earlier been added to the theoretical picture of a black hole. In 1963, the New Zealander Roy Kerr had discovered an exact mathematical description of black holes that is more general than Schwarzschild's solution because it represents the gravitational field around a collapsed object that is rotating, whereas Schwarzschild's equations apply only to a black hole that is not. Thus the so-called Kerr solution, originally regarded (even by Kerr himself) as a mathematical curiosity, has now acquired paramount importance because it describes the distortions of space and time around *any* realistic black hole.

The mathematician Roy Kerr, of the University of Canterbury, New Zealand, who in 1963 discovered the equations that are now believed to describe all black holes in nature.

The realization that the singularity lay *inside* the horizon and not *at* the horizon triggered a drastic shift in the way people thought about collapsed objects and introduced a powerful new metaphor to everyday speech and thought. Before 1968, the hypothetical collapsed remnants of stars were generally called "frozen stars." Physicists had a mental image in which the collapse appeared to slow down and "freeze" at the Schwarzschild radius, even though it was already known that the light from the collapsed object would rapidly become so redshifted that the object's image would quickly disappear from view. It was John Archibald Wheeler who first coined the term "black hole," and the name immediately caught on. Here was something infinitely more mysterious, and perhaps much more sinister as well – a place where *any* form of matter or energy could enter, lose its identity, and be lost forever to the Universe. Is it any wonder that such a concept would resonate with contemporary anxieties, as in the following quote from the June 16, 1980, issue of *Time* magazine, which appears in the *Oxford English Dictionary* definition of "black hole": "To the 1.7 million people added to the jobless rolls in April and May, the U. S. economy may well seem to have . . . been sucked into a black hole." Even if black holes did not exist in nature, they could well have been invented for their metaphorical significance.

Gravitational Equilibrium and Collapse

Black holes represent the ultimate triumph of gravity over all other forces. Gravity is always attractive and, given the chance, will draw everything it can grasp toward a common center. Why, then, is our view of the Universe dominated by structures – stars, clusters of stars, galaxies, clusters of galaxies, and, indeed, the Universe itself – in which gravity seems to be held in check?

The answer lies in the concept of *gravitational equilibrium.* The basic structures in our cosmic environment all engage in a balancing act between gravitational attraction and the dispersive effects of pressure or particle motions. If pressure or particle motions become too large, a system can fly apart, as the entire Universe did some 14 billion years ago during the big bang. Collapse can be staved off as long as a balance is maintained. In the Universe as a whole, gravity currently seems to be losing the competition, overmatched by some mysterious form of pressure – often attributed to a hypothetical "dark energy," though its true nature is completely unknown – that causes the galaxies to fly apart from one another at an accelerating rate. But the systems that form black holes are those in which gravity gets the upper hand, with no recovery possible.

Gravity is not the decisive force in *every* structure, only in the large-scale structures studied by astronomers and cosmologists. It is amazingly weak on the atomic scale: within a single atom, in which negatively charged electrons are electrically bound to a positive nucleus, gravitational forces are weaker than electrical forces by almost 40 powers of 10. In a hydrogen molecule, consisting of two protons neutralized by two (much lighter) electrons, the gravitational binding between the two protons is feebler than their electrical coupling by a factor of 10^{-36}. Gravity is almost undetectable between laboratory-scale bodies as well. In a rock, a person, an asteroid, or even a small planet, the electrical forces between atoms and molecules dominate over their mutual gravitation attraction. But gravitation becomes progressively more significant for larger objects: its relative importance rises a hundredfold for each thousandfold increase of mass.

Why does gravity become powerful only on large scales? In any macroscopic object the positive and negative charges are present in such nearly equal quantities that the electrical forces tend to cancel out. But there is no such cancellation for gravity. Everything has, as it were, the same sign of gravitational charge, and attracts everything else. Gravity is thus the controlling force on sufficiently large scales. How large? Only for masses as large as those of planets can gravity compete with other forces. Indeed, the Solar System's largest planet, Jupiter – made up of 10^{54} atoms (2×10^{27} kilograms, or one-thousandth the mass of the Sun) – is just about the most massive body whose gravity can be balanced by the kind of interatomic forces that operate in solids or liquids. A cold body much larger than Jupiter would actually have a smaller radius because gravity would be strong enough to compress it to an extraordinarily high density.

To comprehend how gravitational equilibrium is *lost*, leading to the formation of black holes, we need to understand first how it is *maintained*. A satellite in orbit around the Earth is continuously falling, but it never actually hits the Earth's surface because its sideways motion carries it just far enough around the circumference to always "miss." Likewise, the motions of all the stars in a galaxy or all the atoms in a star prevent them from being drawn together by gravity to a common center.

The Sun – a typical star – contains a thousand times more mass than Jupiter. If it were cold, gravity would compress it to a million times the density of an ordinary solid – it would be a "white dwarf" star, about the same size as the Earth but 330 000 times more massive. But the Sun is not a white dwarf, because its center has a temperature of about 15 million degrees Celsius, thousands of times hotter even than its

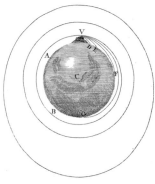

A projectile fired sufficiently fast from a mountaintop would go into orbit because its trajectory would curve downward no more sharply than the Earth's surface. This illustration appeared in *The System of the World*, a "popular" version of the third volume of Newton's *Principia*.

glowing surface. At these high temperatures the atomic nuclei inside the Sun are moving randomly at speeds of hundreds of kilometers per second. It is the pressure of this hot interior – caused by the motions of the fast-moving electrons and nuclei – that counteracts the effect of gravity in all stars like the Sun.

Individual stars are the "atoms" of a galaxy. Galaxies are in some respects simpler systems than their constituent stars. The stars in a galaxy are so widely dispersed, compared to their sizes, that they are unlikely to hit each other; so, to a good approximation, their orbits can be traced independently of one another. Their motions are determined solely by gravity, whereas inside an individual star other physical processes are crucial in determining what happens.

There are about 100 billion stars in our Milky Way Galaxy, which is itself just one of the billions of galaxies within range of modern telescopes. A galaxy is held in equilibrium by a balance between gravity, which tends to pull the stars inward, and the stellar motions, which, if gravity didn't act, would cause the galaxy to fly apart. Disk galaxies, of which our Milky Way is an example, contain a flat "platter" of stars all moving in nearly circular orbits. The star motions are quite regular. In elliptical galaxies the motions are more random, each star tracing out a complicated path under the gravitational influence of all the others. Ellipticals look like big fuzzy balls of stars, sometimes flattened or a bit lopsided. Orbiting around the center of our Galaxy (and others like it) are several hundred *globular clusters* of stars: these are small-scale analogues of elliptical galaxies, generally containing between 100 000 and a few million stars in random orbits.

The amount of motion needed to oppose gravitational collapse can be predicted if one knows how the matter is distributed in space. Imagine that we could dismantle the globular cluster 47 Tucanae, taking each of the million or so stars and carting it off to deep space, so far away from the other stars that their mutual gravitational forces are negligible. To overcome the gravity holding the cluster together we would have to expend a considerable amount of energy, equal to the *gravitational binding energy* of the cluster. The energy required to pull apart a cluster, or any other gravitating object, depends only on the initial distribution of matter and not on the exact procedure used to demolish it. By virtue of their motion, moving stars have the form of energy called *kinetic energy;* it is this energy of motion that counteracts the pull of gravity. In the Newtonian theory of gravity, which is applicable to all systems where the typical motions are much slower than the speed of light, the amount of kinetic energy required for gravitational equilibrium turns out to be exactly equal to the gravitational binding energy. This means, for example, that if we gave the planets in the Solar System

47 Tucanae is one of several hundred globular clusters in orbit around the center of the Milky Way Galaxy. The random motions of its stars balance their mutual gravitational attraction, preventing the cluster from collapsing.

an additional amount of kinetic energy equal to what they already have, they would just be able to break free of the Sun.

Exactly the same balance of energies holds for entire galaxies and clusters of galaxies. This very general result is often called the *virial theorem*. It also applies to stars, except that in the case of a star like the Sun we refer to the kinetic energy as "thermal energy" because a hot gas consists of rapidly moving atoms whose energies are proportional to the temperature. Gravitational equilibrium always requires a balance between the gravitational binding energy and the internal energy opposing gravity.

Gravitational equilibrium does not always last forever. Maintaining the equilibrium in the Sun, for example, requires a central power source to replenish the heat that leaks out through the Sun's surface and is lost to space. Fusion of hydrogen into helium has kept the Sun shining for 4.5 billion years, but this power source is not inexhaustible. The Solar core will run out of hydrogen in another 5 billion years; more massive stars, which shine more brightly, have shorter lifetimes.

When a star faces an energy crisis, gravity compresses its core further. Some other power source may then take over – for instance, fusion of helium into carbon and other elements further up the periodic table. But, at last, all the nuclear fuel that can be tapped will run out. The Sun will cool down to become a white dwarf. Heavier stars collapse further, transforming themselves into more exotic objects.

Even systems composed of stars may not be able to maintain their equilibrium indefinitely. Globular clusters exhibit a gradual form of evolution, which can be observed by comparing the structures of clusters with different masses and ages. A star inside a globular cluster not only responds to the average, "smoothed out" gravitational field of all the other stars, but also feels discrete tugs from its neighbors. The tendency of all these individual encounters is to equalize the average kinetic energies of stars in all parts of the cluster. This sounds as if it should help to preserve the equilibrium, but in fact it has just the opposite effect. The equilibrium is endangered because the gravitational field is weaker in the outer parts of the cluster than it is near the center. If all stars had similar kinetic energies, on average, then some of the stars near the outer edge would be moving fast enough to escape. Thus, globular clusters and other stellar systems can never actually reach a permanent equilibrium; in a sense, they are always slowly *evaporating*.

Let us see how far we can carry the metaphor of evaporation. The loss of the fastest moving stars from the outer parts of the cluster leaves the remaining stars, on average, with less energy. Thus, it would seem that "stellar evaporation" cools the cluster down, just as evaporating water cools the skin. The random motion of stars in the cluster is analogous to the temperature, so we might expect them to move more slowly as the cluster evaporates. However, what actually happens is that gravity pulls the remaining stars more tightly together, so that they end up moving faster than before. The cluster does not cool down as it loses energy – it actually *heats up*!

This unusual (and counterintuitive) property is a general feature of gravitating systems: when they *lose* energy, they get *hotter*. If, for instance, the Sun's nuclear energy supply were turned off, its core would gradually deflate as heat continued to leak out. But as the core shrinks, it has to become even hotter to support itself against the (now stronger) gravitational force. Something similar happens in a star cluster when stars "evaporate" from it. If we consider the structure of a star cluster in more detail, we see that what actually happens is that the central star concentration increases in a runaway fashion. This *gravothermal catastrophe*, first predicted by Vadim Antonov in 1962, can lead to globular clusters with unusually compact, tightly bound cores. More important, it illustrates a general tendency of gravitating systems to develop compact nuclei with very intense gravitational fields.

Globular clusters have been evolving since their formation about 13 billion years ago. In that time, about 20 percent of them have developed dense cores. These cores do not, however, form an exotic object such as a black hole or neutron star in the center of the cluster because the number of stars remaining in the core decreases too rapidly as the cluster evolves. But, aided by additional processes such as stellar collisions and the infall of gas, the dense clusters of stars in the nuclei of galaxies *can* evolve into black holes. Some objects, such as stars, can lose their gravitational equilibrium all at once, suffering a catastrophic collapse or explosion. This is what happens when a massive star runs out of the fuel needed to power thermonuclear reactions. In some cases the object is able to find a new equilibrium, as when the core of a moderately massive star collapses to form a neutron star. But in other cases no equilibrium is possible. The general theory of relativity permits one and only one outcome in this situation – collapse to a black hole.

The Search for Black Holes

Most theories of gravity, not just general relativity, predict that black holes can form, although with slightly different properties depending on the theory. A strong motivation for searching for black holes is that they represent objects where gravity has overwhelmed all other forces, allowing one to test theories of gravitation under the most extreme conditions. But there are other reasons for studying black holes. Stellar-mass black holes interest the astrophysicist because they are in a sense the "ghosts" of dead stars and tell us something about the last stages of stellar evolution. Surprisingly, they give themselves away through the pyrotechnics of flickering X-ray sources. The super-massive black holes in galactic nuclei may be the "trash heaps" of galaxies, or they may be the "seeds" of galaxy formation in the early Universe. In either case, the quasars and jets of gas they produce are the most energetic phenomena known. So great is the energy liberated by the growth of a supermassive black hole that it can heat a volume of gas millions of light-years across; indeed, black holes seem to play a key role in regulating the formation and evolution of galaxies and clusters of galaxies. All these phenomena, from the predictions of general relativity to cosmic jets, offer the astrophysicist a rich source of insights about black holes, and they are the topics of the following chapters.

Black holes are objects that have collapsed, cutting themselves off from the rest of the Universe but leaving a gravitational imprint frozen in the space they have left. A black hole has, in effect, a "surface" resembling a semipermeable membrane. Things can fall in, but the interior is shrouded from view because not even light can escape. No matter how complicated the ancestral object was, once a black hole forms, its appearance from outside has only a few simple characteristics. The thought that complicated and "messy" astrophysical processes could lead to such a simple and "clean" endpoint has endless philosophical appeal for both astronomers and physicists.

To the physicist, gravitational collapse is important also because the singularity that develops inside the hole must be a region where the laws of classical gravitation break down and one needs (just as one does in discussing the initial instants after the big bang) to extend general relativity into the quantum regime to understand what is going on. Thus, black holes are at the frontier of modern physics. Many people have claimed that the paradoxes associated with the singularity within a black hole are as fundamental and far-reaching in their implications as the puzzles connected with blackbody radiation and the stability of bound electron orbits in atoms that confronted physicists at the beginning of the twentieth century and triggered the development of the quantum theory.

Black holes also have an important bearing on our general concepts of space and time because near them spacetime behaves in peculiar and highly "non-intuitive" ways. For instance, time would "stand still" for an observer who, managing to hover or orbit just

"IT'S BLACK, AND IT LOOKS LIKE A HOLE.
I'D SAY IT'S A BLACK HOLE."

outside the horizon, could then see the whole future of the external
Universe in what, to him, was quite a short period. Stranger things might
happen if one ventured inside the horizon. It is therefore hardly surprising that black holes have captured the imagination of the public in a
way that few physical concepts have matched. Even though we cannot
yet predict what would happen very close to the singularity itself,
that limitation need not deter us from the study of black holes. The
"singularity" is hidden within the black hole and cannot directly
affect the external Universe. For this reason, our uncertainty about the
interior of black holes doesn't reduce our confidence in predicting their
astrophysical consequences. As an analogy, there are many mysteries deep inside the atomic nucleus, but this realization doesn't prevent
physicists from calculating the properties of atoms measured by spec-
troscopists and chemists.

The study of black holes has progressed from the cautious
speculation of the mid-twentieth century, to the bold deductions that
followed the discovery of quasars and X-ray binaries, to a rich and
quickly maturing sector of astrophysics today. Questions about the
existence of black holes have given way to studies of their demographics. Astrophysicists are confident that some massive stars do

leave behind black hole remnants, and now seek to determine the masses of these black holes, how fast they spin, and whether they curve spacetime in precisely the manner predicted by general relativity. We are equally confident that the nuclei of most galaxies harbor single, supermassive black holes containing millions of solar masses, and that the formation of these black holes is tightly linked to the formation and evolution of their galactic hosts. In the chapters that follow we lay out what we now know about black holes in the Universe, how we have come to know it, and what remains to be learned.

2 STARS AND THEIR FATES

Ordinary stars like the Sun cannot last forever. The Sun is held in equilibrium by a balance between gravitation and the pressure in its hot interior. Take away that pressure and the Sun would go into free-fall, halving its size in less than an hour. If, on the other hand, gravity were magically switched off, the hot interior would just as suddenly explode and disperse. To us, the Sun looks like a steady beacon in the sky, and in fact its appearance has hardly changed for more than 4 billion years. But the heat of its interior is continuously ebbing. The rays of sunlight you catch on a beach or in a park are plain enough evidence that the Sun is losing energy. And although this heat is now being replenished by nuclear reactions at the Sun's center, these reactions cannot continue indefinitely at their present rate. Eventually the Sun, like any other star, will use up its nuclear fuel, and its finely tuned gravitational equilibrium will fail.

The need for a central power source to replenish the energy lost by a star was recognized in the nineteenth century. If there were no power source at the center, the heat loss would cause the Sun to deflate, at a rate first calculated by Lord Kelvin. The deflation would be far slower than free-fall, because the energy takes a long time to leak out of the Sun's interior – indeed, the process would take about 10 million years. Kelvin knew, however, that 10 million years was much shorter than the   that biologists and geologists had estimated for the age of the Earth and its life forms. Some energy source within the Sun must prolong its life, and Kelvin readily showed that chemical energy would be utterly inadequate. To reconcile his theory with the Earth's estimated age would require, as Kelvin wrote, "some unknown source of energy laid down in the storehouse of creation." Not until the 1930s, when the energetic potential of nuclear fusion was recognized, was there a solution to Kelvin's paradox. The Sun's center is hot enough for hydrogen to fuse into helium at just the rate needed to compensate losses; moreover, the energy release is so high that the supply of hydrogen could keep the Sun shining for several billions of years.

What happens after a star's nuclear fuel is exhausted? This question concerns not only the far distant future. Stars of similar mass to the Sun, but which formed several billion years earlier when our Milky Way was young, have already faced this crisis. Moreover, stars much heavier than the Sun are so much brighter that they use up their fuel more quickly – the most massive ones in only a few million years. Thus, the remnants of dead stars are all around us. These remnants are all much more compact and have much stronger gravity than their precursors.

As their fuel is exhausted, the centers of stars evolve through a sequence of ever more tightly bound gravitational equilibria, punctuated by episodes of contraction. The Sun will become a white dwarf, a slowly

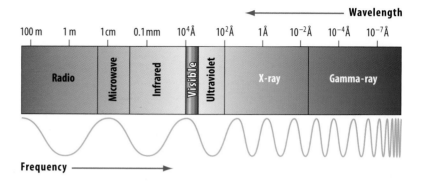

The electromagnetic spectrum. The longest wavelengths and lowest frequencies correspond to photons carrying the smallest amounts of energy. Wavelengths are usually measured in angstroms (1 Å = 10⁻¹⁰ m), frequency in Hertz (1 Hz = 1 cycle per second).

cooling cinder about the size of the Earth, and can remain in this state indefinitely. But stars of much greater mass contract further – they end their lives as neutron stars or black holes. In this chapter, we show how and why these end states are attained and set the stage for discussing, in Chapter 3, the dramatic and unexpected ways in which these seemingly inert remnants have revealed themselves.

Stellar Evolution

Since the 1930s, substantial progress has been made in understanding the structures and life cycles of stars, particularly stars like the Sun. Such stars start their lives by condensing gravitationally from interstellar clouds. At first they contract more or less as Kelvin envisaged, but this process stops when the central temperature becomes high enough for fusion to start in the core. The star then settles down to the so-called main sequence, the longest-lived epoch in a star's life. The main sequence phase lasts until most of the hydrogen in the core (approximately 10 percent of the star's total hydrogen) has been incinerated. During this phase, the power output, size, and surface temperature of the star are all related uniquely to the mass, and to one another.

The most obvious way to discriminate among stars, to a casual observer, is by noting their surface temperatures: temperature determines color (or "spectral type," in astronomers' jargon). Look up on any dark night and you will see that the brighter stars show a distinct range of colors from obvious reds to subtle blue-whites. Though filtered by the response of our eyes, these perceived color differences are very real. The color of light is associated with the wavelength of the corresponding electromagnetic waves. Bluer light has shorter wavelengths; the wavelengths of redder light are longer. Photons, the discrete packets into which light is parsed, each contain a fixed amount of energy that *also* depends on the color. Photons of blue light carry more energy than photons of red light. Since temperature is generally a measure of energy

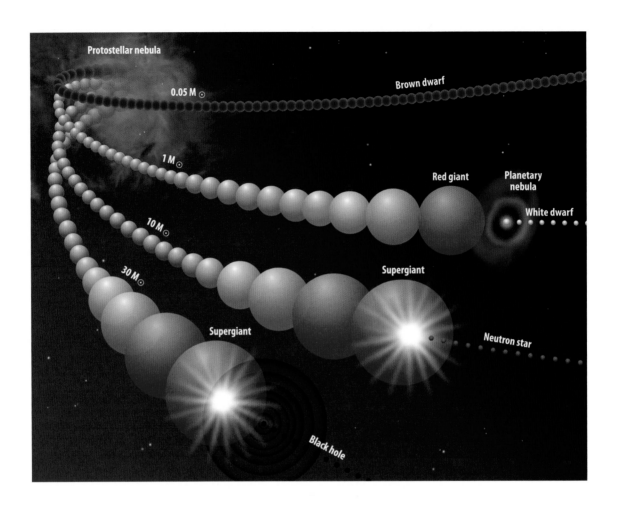

per particle and photons are effectively "particles" of light, it is not surprising that hotter stars are bluer and cooler stars are redder. Simple as this explanation may sound, its ramifications pervade the foundations of modern physics. Why heated objects produce a unique spectrum of radiation that depends only on their temperature was one of the principal puzzles facing physicists at the turn of the twentieth century. Its resolution (due, in part, to Einstein) led to the realization that light behaves both as a wave and as a particle, and thence directly to the quantum theory.

More massive stars are hotter and bigger, but, most important, they are much more luminous. Hence more massive stars burn up their supply of fuel much more rapidly than less massive stars, even though the fuel supply is bigger to start with. The Sun, and stars like it, have enough nuclear fuel to keep shining for 10 billion years. But a 20-solar-

Stars evolve in both structure and appearance as they use up their nuclear fuel. As the nuclear energy supply dwindles, gravity tends to draw the central regions of the star into an ever more tightly bound core. Even during the red giant phase, the bloated stellar envelope conceals a super-dense core not much bigger than the Earth. As illustrated here, a star's ultimate fate depends on its initial mass, the most massive stars ending up as neutron stars or black holes and the less massive stars settling down as white dwarfs. Bodies of very low mass never become hot enough for thermonuclear fusion to start, and exist as brown dwarfs. The demise of a massive star is accompanied by an explosion known as a supernova; stars like the Sun return some of their mass to interstellar space by ejecting a planetary nebula.

mass star, for example, is nearly 10 000 times as luminous as the Sun, and lives only a few million years.

Paradoxically, stars at first become *more luminous* as they use up their store of nuclear fuel. This curious property is related to the fact that gravitating systems "heat up as they cool down," as we saw when we considered the fate of a globular cluster in the last chapter. As nuclear fusion proceeds in the core of the Sun, more and more hydrogen is converted to helium. As the fuel is exhausted, the production of nuclear energy becomes less efficient, and the core shrinks under its own weight. But with this contraction, the gravitational force felt by the atoms in the core increases and the remaining hydrogen atoms speed up. As atomic motion increases, so do the rates of nuclear reactions and, consequently, the power output. Since nuclear reactions can last only as long as the fuel supply, this increase of luminosity cannot last indefinitely.

The changes in the Sun will not be very noticeable for another 5 billion years or so. Thereafter, the solar orb will swell to the enormous dimensions of a red giant star, possibly vaporizing the Earth in the process. But do not imagine that the Sun will be simply dispersing. Deep within the bloated envelope of the red giant Sun, the nuclear furnace will have shrunk to the size of the Earth. During its red giant phase, lasting nearly a billion years, the star burns hydrogen in a shell around its now-helium core. A briefer stage follows, in which helium itself is burnt, after which the star settles down to a quiet demise as a white dwarf.

The story of stellar evolution sketched by astrophysicists is an elaborate one; their theories show that the evolution of a star is determined by the interplay of complicated nuclear reaction chains with subtle aspects of radiation and gravitational physics. How can we check these theories? We cannot perform experiments on stars, we can merely observe them. Moreover, stars are so long-lived compared with astronomers that we only have, in effect, a single snapshot of each one. But as compensation, billions are available for study, and we can check our theories of how stars form and evolve by observing stars of different ages, just as you could infer the life cycle of a tree, even if you had never seen one before, by one day's observation in a forest. The British astronomer William Herschel expressed this point in the eighteenth century: "Is it not almost the same thing, whether we live successively to witness the germination, blooming ... and wither of a plant, or whether a vast number of specimens, selected from every stage through which the plant passes, be brought at once to our view?" Of special interest in this regard are places like the Great Nebula in Orion, where even now new stars, and probably new solar systems, are condensing within glowing gas clouds. Astronomers can also test their theories by studying the stellar populations in globular clusters – swarms of up to a million stars,

held together by their mutual gravity, containing stars of different sizes that can be presumed to have formed at the same time.

White Dwarfs

Gravity does not always win in the end; the Sun's fate as a white dwarf promises to be a stable one. White dwarfs are a kind of star that can remain in gravitational equilibrium indefinitely, without any need of nuclear reactions or other power supply. Like ordinary stars, white dwarfs are supported against gravitational collapse by the random motions of the particles of which they are composed. But, unlike those in a normal star, these motions are *not* related to the temperature of the gas in the interior. So, as white dwarfs radiate away their internal thermal energy and cool down, they do not lose pressure support and contract.

The kind of pressure that supports a white dwarf is called *degeneracy pressure*, and it arises from a quantum mechanical effect called the *Pauli exclusion principle*: two identical particles cannot have the same position and momentum. The principle applies to the basic building blocks of atoms – protons, electrons, and neutrons. Such particles are collectively called fermions, after the great Italian physicist Enrico Fermi. What they all have in common is that they are endowed with precisely one-half (or, more precisely, a half-integer) unit of quantum mechanical angular momentum, or "spin." The spin is an intrinsic property of each kind of particle, and there is nothing one can do to change it. Despite the esoteric sound of the concept, "half-integer spin" has a very concrete effect on the collective behavior of groups of fermions: it prevents them from being packed together too closely. More precisely, it prevents any two of them from occupying the same "quantum state," meaning that they cannot both lie very close together in space *and* have similar velocities. When fermions are packed closely together, they cannot all be slow moving. The exclusion principle forces most of them to have high velocities. The more you compressed them, the faster they would move. These motions give rise to a pressure that resists compression, over and above any pressure the particles might exert as a result of their thermal energy.

When most of the particle motion in a gas is the result of this quantum-mechanical resistance to compression, the gas is called *degenerate*. The random motions associated with degeneracy create a pressure that can support a star against gravitational collapse, just as the more familiar thermal motions create the pressure that supports the Sun. A star can become a white dwarf when its constituent electrons

In 1930, Subrahmanyan Chandrasekhar calculated that white dwarfs more massive than 1.4 suns would collapse under their own weight, paving the way for the theoretical prediction of neutron stars and black holes. In an extraordinarily long and prolific career (mainly at the University of Chicago) he wrote major books on many topics in astrophysics, including *The Mathematical Theory of Black Holes*, published in 1983. He published a detailed commentary on Newton's *Principia* shortly before his death in 1995.

become degenerate; this happens when the star is compressed to a density about a million times that of water. White dwarfs are roughly the size of the Earth, but with a gravitational field about a million times larger!

There is, however, an upper limit on the mass of a white dwarf. If you examined a sequence of white dwarfs of increasing mass, you would find that the degeneracy-induced random speed of the electrons increases with the mass of the dwarf, approaching the speed of light for masses just above that of the Sun. In 1930, the young theorist Subrahmanyan Chandrasekhar (known to his colleagues as Chandra) pondered this problem during the long voyage from India to England, where he had enrolled as a graduate student, and reached an astonishing conclusion. He found that no gravitational equilibrium could exist for white dwarfs with masses above about 1.4 solar masses. At higher masses, the degeneracy pressure could never balance gravity, no matter how much you compressed the star. Chandra realized that forcing the motion of the electrons to speeds close to the speed of light would weaken the resiliency of the degenerate gas, its ability to resist compression; consequently, a white dwarf with a mass above the "Chandrasekhar limit" would go into gravitational free-fall and collapse in about one second!

The implications of Chandrasekhar's discovery were disturbing. What would eventually happen to stars of mass greater than 1.4 solar masses once they had exhausted their nuclear energy? Some suggested that a massive star in the course of its evolution would lose most of its mass, so that no matter how massive it started out it would eventually settle down as a white dwarf of mass below the Chandrasekhar limit. We now know that massive stars do lose a lot of mass during their lifetimes, through powerful stellar winds, and in this way, stars moderately more massive than 1.4 solar masses probably end up as white dwarfs. But there is no evidence to suggest that stars of much greater mass can avoid Chandra's mass limit. Another possibility is that the outer layers may be ejected explosively in a supernova event at the end of the star's lifetime, leaving a neutron star. But theory sets a limit to the mass of a neutron star. The exact limit is a bit uncertain, but it certainly lies below 3 solar masses. The third possibility is unchecked collapse to a black hole.

In 1930 neutron stars had not yet been conceived, and so unpalatable was the third possibility that many scientists at the time simply refused to accept Chandra's prediction at face value. The famous theorist Sir Arthur Eddington led a public attack, rejecting the fundamental physical basis of Chandra's arguments. "Various accidents may intervene to save the star, but I want more protection than that," he proclaimed at a meeting of the Royal Astronomical Society in 1935.

"I think there ought to be a law of Nature to prevent a star from behaving in this absurd way!" Eddington was ultimately proved wrong, but Chandra, stung by the ferocity of the attack, left British academia permanently for the more welcoming environment of the University of Chicago. Chandra went on to a brilliant career as a pioneer in many other areas of astrophysics, but he was so discouraged by the white dwarf episode that it was decades before he could bring himself to tackle the problem of gravitational collapse once again – although he did so in a characteristically exhaustive fashion with his treatise on the mathematical theory of black holes, published in 1983.

Massive Stars and Supernovae

Stars heavier than the Sun evolve in an increasingly complicated and dramatic way. Most end up (just as the Sun will) as white dwarfs composed mainly of carbon and oxygen (although there may be a thin layer of hydrogen and/or helium remaining on the surface). Their interiors never become hot enough to trigger the nuclear fusion of carbon and oxygen into heavier nuclei. Whereas electrical or gravitational forces can act over arbitrarily large distances, the forces that cause nuclear reactions operate only over distances less than about 10^{-13} centimeters, or about one hundred-thousandth the size of an atom. Therefore, to trigger a nuclear fusion reaction it is first necessary to bring the reactants very close together. (However, one need not bring them as close as 10^{-13} centimeters, because a quantum mechanical effect called tunneling makes the particles behave as if they were smeared out over a larger region in space.) Although a nuclear fusion reaction requires close encounters between nuclei, the protons in a nucleus electrically repel the protons in other nuclei (since like charges repel). This electrical repulsion tries to prevent the particles from getting too close together. Heavier atomic nuclei contain a larger number of protons and therefore repel each other electrically with more force. High relative velocity – or, equivalently, high temperature – is needed to overcome the repulsion.

Massive stars are powered in later life by a sequence of nuclear reactions involving heavier and heavier elements. As each nuclear fuel is exhausted – hydrogen fused into helium, then helium into carbon and oxygen, etc. – the inner part of the star contracts, becoming even hotter, until a new set of fusion reactions can take the atoms further up the periodic table. Computer models of stellar evolution indicate that the Sun will probably burn much of its core to carbon before it settles down as a white dwarf. But for stars more massive than 5 to 8 suns (the borderline is uncertain), this process would proceed all the way up to iron. At every stage up to this point, the creation of heavier atomic

The interior of a massive star goes through a series of nuclear-burning stages in which increasingly heavy elements are created until the core consists of pure iron. Outside the core the star has an "onionskin" structure in which the chemical composition of each layer is determined by the maximum temperature it reached during the star's life. No nuclear energy can be extracted from iron, and if the core is more massive than 1.4 suns it will collapse, forming a neutron star or black hole. In this illustration showing the structure of a 20-solar-mass star, the collapse of the core blows off the outer layers, producing a supernova explosion.

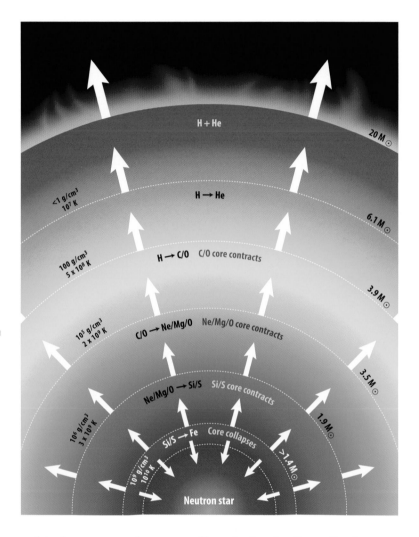

nuclei releases energy that staves off gravitational collapse. But there are no nuclear reactions that can release energy from iron; iron is the end of the nuclear road for a star.

What happens next is one of the most spectacular events known in astronomy. A sufficiently massive star (probably any star above about 8 solar masses) develops an iron core that weighs more than 1.4 suns. In other words, the core is too massive to reach equilibrium as a white dwarf – it exceeds Chandrasekhar's limit. Since there are no nuclear reactions that can extract energy from iron, the supply of fuel is shut off and the core suffers sudden and catastrophic collapse, reaching the density of an atomic nucleus (about a million billion times the density of water!) in a fraction of a second. During this collapse, the nuclear reactions that have built up the iron core are undone – the iron nuclei are broken apart into smaller and smaller pieces until a soup of subatomic particles is all that remains.

If a sequence of nuclear or chemical reactions has released a certain amount of energy, then reversing those reactions requires the introduction of an equivalent amount of energy. Where does this energy come from? Gravity, of course! The collapse of the core releases an amount of energy equal to about *10 times* the energy output of the star during its entire nuclear burning lifetime. To put this another way: over the lifetime of a massive star, gravity, not nuclear reactions, provides most of the energy, and does so almost entirely within the last second of the star's life.

The dismantling of the nuclear fusion products does not stop with the breaking apart of iron into subatomic particles. The density in the collapsing core becomes so great that the protons and electrons among these particles are fused together to form neutrons, electrically neutral subatomic particles with about the same mass as protons. Normally, neutrons are stable only when they are bound to protons inside atomic nuclei. An isolated neutron will decay into a proton, an electron, and a third particle called an antineutrino in about 10 minutes. But in the dense core, the existing electrons are already so tightly packed that the Pauli exclusion principle does not allow any room for additional electrons to be created by neutron decay. The core of the collapsing star becomes a sort of gigantic atomic nucleus containing 10^{57} neutrons, and will (if it is not too massive) evolve into a neutron star. The force that supports a neutron star against collapse is the degeneracy pressure set up among the neutrons, which also belong to the fermion class of particles.

If gravity has released 10 times more energy than needed to dismantle the atomic nuclei, where does the surplus 90 percent of the energy go? About 90 percent of that surplus is carried away by subatomic particles called neutrinos, which are released during the formation of the neutrons. Neutrinos are particularly elusive particles that hardly interact with ordinary matter; most of them pass through the envelope of the dying star without doing much damage. The other 10 percent, however, triggers an explosion that blows off the outer layers of the star. These explosions are called supernovae, rare events that occur on average only once or twice per century in our Galaxy.

One such event created the Crab Nebula in the constellation Taurus. Although the light from this explosion reached the Earth in the year A.D. 1054, we fortunately have records of the supernova's appearance from the Orient, particularly China. In the fifth month of the first year of the Chih-Ho reign, the Emperor received a visit from Yang Wei-Te, the Chief Calendrical Computer, equivalent perhaps to the British Astronomer Royal. Yang is recorded to have said to the Emperor, "Prostrating myself before your Majesty, I have observed the appearance of a guest star. On the star there was a slightly iridescent

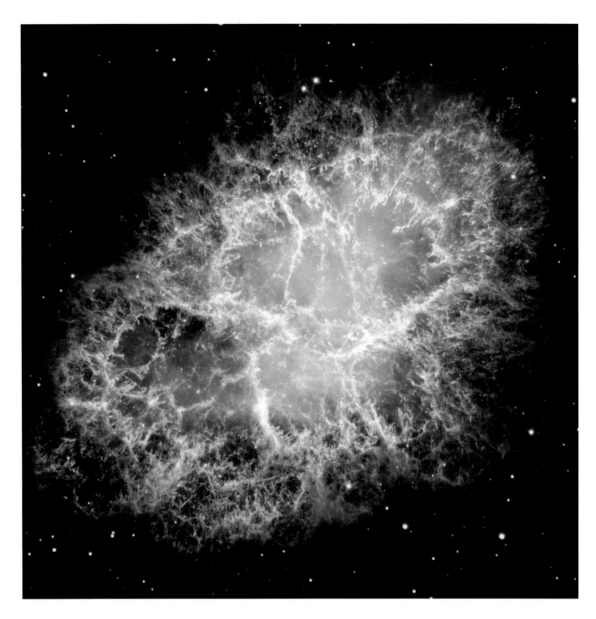

yellow color. Respectfully I have prognosticated, and the result said, the guest star does not infringe upon Aldebaran. This shows that a Plentiful One is Lord, and the country has a Great Worthy. I request that this prognostication be given to the Bureau of Historiography to be preserved." Now, nearly a thousand years later, we see the expanding debris from the explosion witnessed by Yang Wei-Te. Supernova explosions signify a violent endpoint of stellar evolution. Astronomers now distinguish several different types of supernovae, but the most

The Crab Nebula is the remnant of the supernova explosion observed in A.D. 1054. In this mosaic of images from the *Hubble Space Telescope*, the red and green filaments are shredded debris of the star's envelope; the diffuse bluish light comes from electrons gyrating at close to the speed of light in a magnetic field that pervades the nebula.

common mechanism (the one that gave rise to the Crab Nebula) comes about when a star much heavier than the Sun, too massive to leave behind a white dwarf, exhausts its nuclear fuel. The star then faces an energy crisis. Its core catastrophically implodes, and the outer layers are blown off.

Supernovae are bright enough to be observed at great distances. Thus, although they are rare events in any one galaxy, by watching the behaviors of many supernovae discovered in other galaxies, astronomers have pieced together a systematic description of how their spectra and luminosities change with time following the initial explosion. But the large distances of typical supernovae have handicapped observers, especially when the supernova fades out of view after a few months. Thus the discovery, on February 23, 1987, of a "nearby" supernova was a wonderful stroke of luck for astronomers. Supernova 1987A exploded a mere 170 000 light-years away in the Large Magellanic Cloud, a satellite galaxy of the Milky Way. Supernova 1987A is probably the first supernova for which the star was observed *before* it exploded. The doomed star did not exactly fit theoretical expectations, being blue rather than red, but this anomaly (which is now consistent with revised theory) is minor compared with several impressive verifications of our general theoretical picture of supernovae. The star was massive – about 20 solar masses – and its mass fits the picture of a core collapse. But more important, two deep underground neutrino detectors – Kamiokande II in Japan and the IMB detector in Cleveland, Ohio – each captured a handful of neutrinos produced during the collapse and "neutronization" of the stellar core. Having traveled all the way through the stellar envelope and intervening space, these neutrinos reached the Earth in a burst lasting less than a minute, arriving several hours before optical telescopes could observe the shock wave erupt through the surface of the star. (However, the data from the neutrino detectors were not analyzed until after the supernova had been discovered by the traditional optical means.)

The remarkable observation of neutrinos provides the first direct confirmation of the core collapse theory for supernovae. In fact, it was the first detection of *any* neutrinos from an astronomical object other than the Sun. Neutrinos from supernovae are not rare, but they are difficult to detect because they hardly interact with any form of matter, including the matter in a neutrino detector. Just as most of the neutrinos produced during a supernova escape even the dense core of the collapsing star, so did virtually all of the neutrinos reaching Earth from SN 1987A pass through unnoticed, neutrino detectors and all. In fact, about 100 trillion neutrinos from 1987A passed through each person on the planet. Neutrino detectors put up a considerably larger obstacle than a person, and caught only 10 apiece. Some astronomical wag has estimated that a half-dozen or so of the five billion earthlings might have

(a)

(b)

"Before" (a) and "after" (b) pictures of Supernova 1987A in the Large Magellanic Cloud. The supernova is at the lower right-hand extremity of the galaxy. Using more detailed images, astronomers have determined exactly which star exploded.

had a 1987A neutrino actually hit a molecule inside an eyeball, perhaps creating a noticeable flash. Please do not contact us if you believe you have had such an experience!

Most of what we know about Supernova 1987A comes not from neutrino detectors but from spectral analysis of the supernova's radiation, as explained in the box on pages 38–39.

The detection of spectral lines of cobalt in the aftermath of Supernova 1987A was an important confirmation of supernova theory. In the moments before and during the explosion, the stellar core becomes so hot, and the nuclear reactions so rapid, that some unstable nuclei heavier than iron are synthesized. Among these are cobalt and nickel. The gradual decline in a supernova's light, lasting several months, is attributed to the radioactive decay of unstable forms (isotopes) of cobalt and nickel, produced behind the violent shock wave at the base of the expanding envelope. Supernova 1987A gave us the first direct evidence that cobalt was indeed present in the expected amounts. Unprecedented monitoring of the expanding remnant, at wavelengths ranging from radio waves to gamma rays, has allowed us to "reconstruct" the structure of the stellar envelope and the dynamics of the explosion. Observers also expected to detect signs of a neutron star remnant peeking through – but its whereabouts remain a mystery, leading some to suggest that SN 1987A formed a black hole instead.

It may seem strange that the catastrophic collapse of the inner part of a star should lead to the violent explosion of the outer part. Indeed, it has proved quite difficult to produce a satisfactory theory to explain the details of supernova explosions. There seem to be a number of effects that conspire to throw off the star's envelope at high speed. One is the so-called core-bounce mechanism. When the core approaches

When a star explodes as a supernova, its outer layers are blown off at up to 20 000 kilometers per second. Its brightness flares up within a few days, and then gradually fades. To optical astronomers, a young supernova looks like a point of light. However, the techniques of very-long-baseline interferometry (see Chapter 6) allow radio astronomers to obtain sharp enough images to reveal the actual expansion. This series of pictures shows Supernova 1993J, in the galaxy M81, during the first seven years after the explosion.

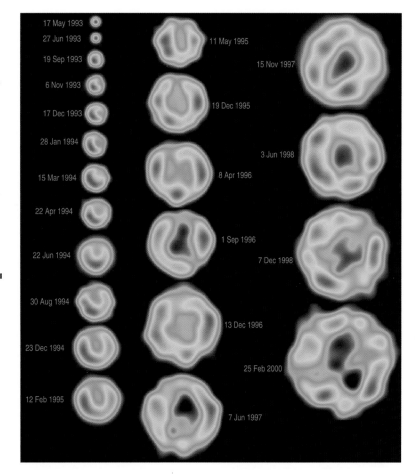

the density of an atomic nucleus, the now-degenerate neutrons become very "stiff," resisting further compression. This sudden increase in pressure is enough to halt the collapse. But the core is contracting so rapidly that instead of screeching to a halt it overshoots and bounces back a little. The idea of the core-bounce mechanism is that the bouncing core runs into the still-infalling envelope, giving it a kick that sends it flying outward. Unfortunately, computer models of this process indicate that it falls slightly short, and that the shock wave driving the explosion "stalls" inside the envelope. A second contributory mechanism is thought to come from the large outpouring of neutrinos from the very hot bouncing core. Although neutrinos hardly interact with matter, there are so many of them that they might give the envelope enough of an extra "push" to re-energize the shock wave and blow off the envelope. Neutrinos do help to trigger an explosion, according to some computer simulations; however, the mechanism seems to require special, rather restrictive conditions in order to work.

Spectroscopy

An atom consists of a positively charged nucleus surrounded by orbiting electrons. Electrons on smaller, closer-in orbits have lower energy than electrons on more distant orbits. An electron can lose energy and drop into a lower orbit by emitting radiation, but if it radiated according to the nineteenth-century "classical" theory developed by James Clerk Maxwell, it would spiral right into the nucleus. Modern-day quantum theory states that this is impossible. According to quantum theory, electrons cannot be concentrated arbitrarily close to the nucleus; instead, there is a well-defined orbit of lowest energy called the ground state. Higher-energy orbits, or "excited" states, are also discrete. So an electron that has become electrically bound to a nucleus, instead of spiraling in gradually, changes its energy in discrete steps and reaches a definite lowest state where it can radiate no more.

Radiation can be viewed either as a wave or as consisting of "quanta," or particles, called photons. A photon carries an energy equal to the frequency (equivalently, the speed of light divided by the wavelength) of the light it represents multiplied by Planck's constant. It requires bigger "jumps" of energy to emit blue light than red light, because blue light is carried in more energetic quanta. Ultraviolet and X-ray emissions require more energy still.

The simplest atom is hydrogen, with just one electron; its energy states correspond to a simple ladder. In very hot gases, as are found in stars and galactic nuclei, collisions and other processes can dislodge electrons from atoms, creating what is called a "plasma" composed of atomic nuclei and free electrons. Whenever an electron is recaptured by a nucleus and initially lodges in an excited state, one or more photons (each with the frequency characteristic of the particular energy jump) will be emitted as it cascades down to the ground state.

A hot gas of hydrogen emits a series of discrete frequencies, which are in simple mathematical ratios. The strongest emission of all, Lyman alpha, corresponds to a

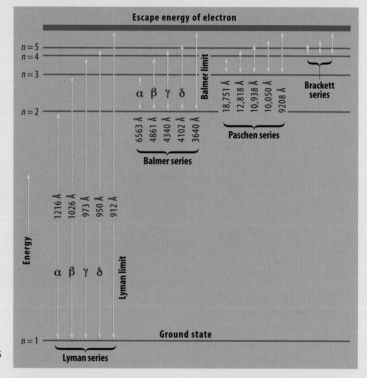

A diagrammatic representation of the permitted energy levels of an electron in a hydrogen atom. Transitions from lower to upper levels absorb energy in discrete amounts, while those from upper to lower levels release energy in the same amounts. The energy released or absorbed in a transition is proportional to the length of the arrow connecting the levels. There is a definite relationship between the energy of a "quantum" of radiation (or "photon") and its wavelength. The spectrum of the light from hydrogen consequently displays a set of lines at well-defined wavelengths. Other atoms produce equally distinctive (though generally more complicated) sets of spectral lines.

transition from energy state $n = 2$ to the ground state $n = 1$. Other transitions ending at $n = 1$ (3 to 1, 4 to 1, 5 to 1, etc.), are known as Lyman beta, gamma, and so on; there is another set of frequencies corresponding to transitions that end up at $n = 2$ rather than $n = 1$; these are the Balmer series. Other series end at $n = 3$, $n = 4$, and so on.

The variation of radiation intensity with wavelength is called the *spectrum*. A *spectrograph* displays a spectrum by spreading the wavelengths in a band across a photographic plate or other detector, from short wavelengths at one end to long wavelengths at the other. At the wavelengths where atomic transitions are producing a lot of emission, the spectrograph shows sharp, bright stripes called *emission lines*. Atoms of each element in the periodic table emit radiation with a characteristic set of discrete frequencies, producing a distinctive set of lines in the spectrum that are the "fingerprint" of that particular element. We can also tell whether an atom has lost some of its electrons – that is, become *ionized* – by measuring the pattern of spectral lines. By measuring the *relative* intensities of different emission lines, we can often deduce the temperature and density of the gas.

The spectrum emitted by a hot, dilute gas consists mainly of emission lines. But a dense or opaque gas, or a heated liquid or solid (such as the filament of an incandescent light bulb), emits radiation at all wavelengths. This kind of continuous spectrum is called *blackbody radiation*. The variation of intensity with wavelength depends only on the temperature of the radiating material, and not at all on its composition. Inside a star, for instance, the radiation would have an almost exact blackbody spectrum.

A pure blackbody spectrum has no features that can reveal the nature of the material emitting the radiation. We can nevertheless learn about the composition of stars by measuring their spectra. The Sun, for instance, has a surface temperature of about 5800 degrees Kelvin. But the surface is not completely "sharp," and successive layers near the surface have slightly different temperatures. When a hot continuous spectrum passes through cooler gas along the line of sight, the inverse of emission – absorption – can occur. An electron can be excited to a higher level by photons of the right energy. This process happens in the outer layers of stars, where a cooler gaseous atmosphere overlays the hot gas of the stellar disk. The light from the star is therefore depleted at the frequencies that have been absorbed. Spectra then show dark rather than bright lines at the wavelengths characteristic of the atoms and ions in the star's outer layers.

Spectrographs can be used to probe the compositions, temperatures, and densities of astronomical bodies. A cloud of hot, rarefied gas produces a spectrum of emission lines, whose wavelengths reveal the cloud's chemical composition and level of ionization. The relative intensities of different lines can give indirect information about the temperature and density of the radiating gas. A dense or opaque gas produces a continuous spectrum, which carries no information about the composition of the source but gives us a direct measurement of its temperature. If a source of continuous radiation is viewed through a cooler intervening cloud of gas, one sees absorption lines superimposed on the continuous spectrum.

(a) **(b)**

(c) **(d)**

While we seem to have the basic theoretical picture in place, we have not yet reached the point where we can simulate a supernova explosion inside a computer. Early calculations assumed (for computational simplicity) that the star was exactly spherical, but this proved to be too much of a simplification. Supernova explosions are extremely messy, turbulent events. They are sensitive to many details, such as asymmetries of the collapsing core and envelope, chemical inhomogeneities, and the effects of rotation and magnetic fields. Supercomputers are increasingly able to handle these complications, but it will still be some time before most of the details are understood. Nevertheless, Supernova 1987A has been a great ego-builder, showing supernova theorists that they have probably been on the right track.

Looking into the heart of a supernova. Four snapshots from a computer simulation show the vigorous boiling of the neutrino-heated, convective region around the nascent neutron star. Buoyant bubbles of hot matter moving outwards appear bright red and yellow. These are bounded by the shock wave, which expands outwards, disrupting the star. The images show the structure at (a) 0.1, (b) 0.2, (c) 0.3, and (d) 0.5 seconds after the shock is born. At these times, the shock has an average radius of about 200, 300, 500, and 2000 km, respectively. Simulations like these show that supernova explosions are likely to be highly asymmetric. Despite their complexity, such simulations are still a highly simplified representation of reality.

Supernovae may seem like very distant and remote events, but they are crucial for our very existence. Throughout our Milky Way, gas is being converted into stars, and material from stars recycled back to gas, enriched in newly synthesized chemical elements through supernova explosions. Complex chemical elements are built from hydrogen via the nuclear reactions that provide the power source in the cores of ordinary stars. Although the elements in the core are destroyed during the collapse just before a supernova explosion, plenty remain in the star's outer shells to be scattered through space by the force of the explosion, and the heaviest elements are created in the explosion itself as the shock wave ignites nuclear reactions in the star's outer envelope. Only by studying the births of stars, and their explosive deaths, can we tackle such an everyday question as where the atoms we are made of came from.

The abundances of the elements are quite well known: they can be determined in the Solar System by direct measurement and inferred spectroscopically in stars, nebulae, and even in other galaxies. The relative abundances display many similarities from place to place, and the pattern is now reasonably well understood. Most elements heavier than helium are made in stars, and blown off either in stellar winds or by a final explosion. We understand why oxygen is common, but gold and uranium are rare. All the carbon, oxygen, and iron on the Earth could have been manufactured in stars that exhausted their fuel supply before the Sun formed. The Solar System then condensed from gas contaminated by debris ejected from these early stars.

Poo-Bah (in Gilbert and Sullivan's *Mikado)* traced his ancestry back to a "protoplasmic primordial atomic globule," but the astronomer goes back further still. Each atom on Earth can be traced back to nuclear reactions inside stars that died before the Solar System formed. An oxygen atom, forged in the core of a massive star and ejected when the star explodes as a supernova, may spend hundreds of millions of years wandering in interstellar space. It may then be once again in a stellar interior, where it can be transmuted into still heavier elements. Or it may find itself out on the boundary of a new solar system, incorporated into the fabric of a planet, and maybe eventually into a living cell. We are literally the ashes of long-dead stars.

Neutron Stars

The ejecta of supernova explosions contain all the familiar chemical elements of which we are made, but the remnant left behind could not be more exotic or alien to our everyday experiences. Seventy-five years ago, the astronomers Walter Baade and Fritz Zwicky made a very bold prognostication, one more physical (though no less speculative!) than Yang Wei-Te's. In a 1934 paper they wrote, "With all reserve we advance the view that a supernova represents the transition of an ordinary star into a neutron star, consisting mainly of neutrons." They conjectured that a supernova explosion was driven by the gravitational energy impulsively released when the star's core collapsed, and that a tiny cinder should remain: a star squeezed to a size of 10 kilometers, and having an internal density comparable to that of an atomic nucleus, 10^{15} times higher than an ordinary solid and a billion times higher even than a white dwarf.

Although theorists, among them J. Robert Oppenheimer and George Volkoff, explored the physics of neutron stars in the late 1930s, their existence remained conjectural until 1968. Jocelyn Bell, then a graduate student at Cambridge University's Mullard Radio Astronomy Observatory, and her thesis advisor, Anthony Hewish, were surveying the sky with a new radio telescope when they found a source emitting regular pulses of radio energy every 1.3 seconds. They soon found some others, each "ticking" with a distinctive regular period. The nature of these sources – later called pulsars – was baffling. They clearly had to be very compact in order to vary so rapidly. One idea, favored by Franco Pacini and Thomas Gold, was that the sources were spinning neutron stars.

Only a year later, John Cocke, Michael Disney, and Don Taylor at the University of Arizona discovered that the ordinary-looking little star in the middle of the Crab Nebula was actually flashing on and off 30 times a second. Nothing much larger than a neutron star could pulse or spin this frequently – even a white dwarf would be too large. So the discovery of the pulsar in the Crab Nebula clinched the case for pulsars being spinning neutron stars whose strong magnetic fields produce a kind of lighthouse beam of radiation that swings through our line of sight once every revolution. Since then, nearly two thousand pulsars have been discovered in the Milky Way Galaxy. Most of them were formed within the last 10 million years and are rather close by; there is every reason to believe that there are at least 100 million neutron stars in our Galaxy, one for practically every supernova that has occurred.

The discovery of a fast-spinning pulsar solved the puzzle of why the Crab Nebula still glows so brightly, more than nine hundred years after the explosion that created it. The glow of a typical nebula, like the Orion Nebula, comes from turbulent shocked gas and from atoms bathed in ultraviolet radiation from hot young stars. The nebula pervading the Pleiades star cluster in Taurus glows blue because tiny dust particles scatter the light of the stars. But the diffuse blue light from the Crab Nebula has a different origin. It does not come from ordinary hot gas; rather, it is synchrotron radiation, emitted by electrons or their anti-matter counterparts, positrons, moving at very close to the speed of light through a magnetic field. It is straightforward to calculate how quickly these ultra-energetic particles (with energies a million times higher than their mass-energy equivalent, mc^2) will radiate their energy away – they

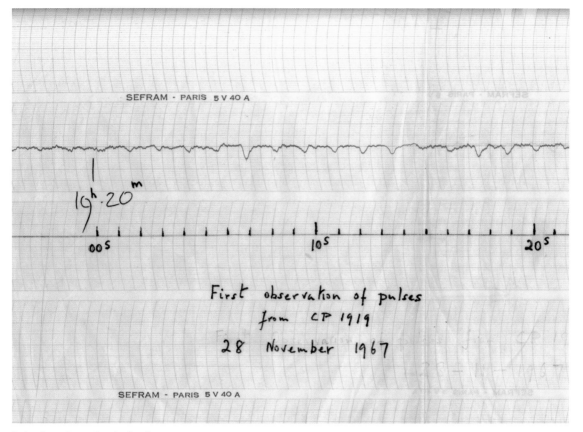

SEFRAM · PARIS 5 V 40 A

$10^h.20^m$

00ˢ   10ˢ   20ˢ

First observation of pulses
from CP 1919
28 November 1967

SEFRAM · PARIS 5 V 40 A

Strip-chart data taken by Jocelyn Bell in 1967 record the first detection of a pulsar.

radiate so quickly that they cannot have retained their energy since the explosion occurred in A.D. 1054. The tiny spinning star must be acting like a particle accelerator, continuously regenerating the fast electrons that produce the blue light. In effect, the star transfers some of its rotational energy to the electrons. The timing of the pulses shows that the star is slowing down, and the rate at which it is losing its rotational energy is more than enough to keep the nebula shining.

Through neutron stars, the cosmos offers us a laboratory where we can study how material behaves under conditions far more extreme than can be produced on Earth. Basically, a neutron star is an analogue of a white dwarf in which degeneracy pressure among neutrons, rather than electrons, provides the force that opposes gravity. We cannot determine the upper limit for the mass of a neutron star by simple analogy with the Chandrasekhar-limit argument for white dwarfs because nuclear forces figure importantly in the gravitational equilibrium, and our understanding of the behavior of nuclear matter at such high densities is incomplete. Still, we can be quite confident, from general theoretical considerations, that neutron stars cannot exist with masses exceeding 3 solar masses. At

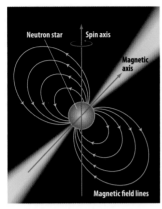

Neutron stars have ultra-strong magnetic fields. If the stars are spinning, these fields generate strong electric currents, as in a dynamo, and accelerate electrons. These electrons emit an intense beam of radiation from each magnetic pole. Because the magnetic field is misaligned with the star's spin axis, these beams sweep around the sky once per revolution.

least theory does tell us quite a bit about what the interior of a neutron star should be like. It should have a solid crust, a transitional region that is a liquid – indeed, a so-called superfluid having almost no viscosity – and a core about which little is known in detail.

How can we test whether these properties really characterize the inside of a neutron star? One cannot, of course, go drilling below the surface of a neutron star, as a geologist can drill below the surface of the Earth. At first sight, it might seem hopeless to check out ideas about the inside of a neutron star by looking at one thousands of light-years away, but amazingly enough we can.

Pulsars can be timed with microsecond precision. Though they are on average slowing down, there are occasional "glitches," when they suddenly speed up by a tiny amount. There is one straightforward reason to expect glitches. A spinning object is flattened as it rotates by inertial forces (often called "centrifugal force," although some physics teachers excoriate that term), creating an equatorial bulge. As a pulsar slows down, these forces decrease and the bulge tries to shrink. If the neutron star were a fluid, the equatorial bulge would diminish continuously. But if the star has a solid crust, a continuous readjustment is impossible. Stress will build up, and when it exceeds a certain limit the crust will crack. Thus the readjustment takes place in jerks, causing sudden increases in the spin rate, the size and frequency of which tell us how thick and rigid the crust is. These changes in spin rate can be measured with such precision that a decrease of a few millionths of a meter in equatorial radius would show up clearly in the timing data on pulsars thousands of light-years from us.

The behavior of glitches has been studied in detail, and it is now believed that the most common ones result from a different effect: slippage between the solid crust and the fluid core. The drag that slows down a pulsar's rotation is applied to the crust and does not directly slow down the liquid core. A differential spin rate therefore develops between crust and core. The slowdown of the crust is communicated inward by a frictional force. This force doesn't act smoothly, however: instead, it operates in jerks rather like a car's worn-out clutch and causes the crust to spin up whenever the friction suddenly increases.

The precise study of how the spin rate changes is just one of the ways we can learn about the insides of neutron stars. Another technique, made possible by sensitive X-ray telescopes in space, provides a direct determination of the relationship between a neutron star's mass and its radius by measuring the spectrum of radiation from the glowing surface. These kinds of measurements can distinguish between conservative models for the composition of neutron star cores and more exotic possibilities such as "strange quark matter."

A slice through a neutron star would in some respects look rather like a slice through the Earth, with a crust, a liquid interior, and

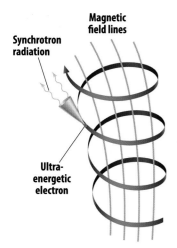

Synchrotron radiation

Magnetic field lines

Ultra-energetic electron

Synchrotron radiation, produced when electrons move through a magnetic field at close to the speed of light, is responsible for the bluish glow of the Crab Nebula. The electrons trace out helical trajectories centered on the magnetic lines of force. Synchrotron emission from each electron is concentrated in a narrow cone around its direction of motion.

maybe a solid core. The theory is detailed enough to predict that the star should deviate from a pure neutron composition at various depths. The outer crust would consist mainly of iron, but at higher densities the main constituents would be progressively more exotic neutron-rich nuclei such as nuclei of nickel, germanium, and krypton, arranged in a crystalline lattice. Many of these nuclei would be unstable in the laboratory, as are isolated neutrons: they undergo beta decay, emitting an electron and antineutrino. In a neutron star, however, this decay is inhibited according to the Pauli exclusion principle, because there are no unoccupied quantum states into which the electrons can go.

Free neutrons begin to appear in significant numbers at densities greater than about 400 billion grams per cubic centimeter – the so-called "neutron drip" point. Digging still deeper below the crust one reaches densities approaching 10^{14} grams per cubic centimeter. Under these conditions the matter is mainly in the form of free neutrons. Moreover, the neutrons display the properties of a superfluid analogous to those displayed by terrestrial helium at temperatures very close to absolute zero. Superfluids have very unusual properties, including negligible internal viscosity. In fact, a vortex once set spinning within a superfluid will spin indefinitely. Within this fluid formed of free neutrons, a small fraction of the material remains in the form of protons and electrons. The proton component (and perhaps the electron component as well) would behave as an electrical superconductor, that is, a material with zero electrical resistance. The pressure exerted by "nuclear matter" at these densities is not exactly known; there is consequently some uncertainty in the internal structure and radius of a neutron star. Right in the core, where the density may attain values of several times 10^{14} grams per cubic centimeter, the character of the material is even more uncertain, and perhaps stranger still. One possibility, still highly controversial, is that the mix of elementary particles in a neutron star's core could condense into "strange quark nuggets" – a solid made from crumbled bits of protons and neutrons, as well as other elementary particles.

Neutron stars also appear to resemble the Earth in that they possess magnetic fields, but the magnitudes of these fields are vastly larger. The surface magnetic strength of a pulsar is typically several trillion times stronger than that of the Earth and a million times stronger than the strongest laboratory fields. In a subclass of neutron stars – the "magnetars" – the magnetic field is a thousand times stronger still. For comparison, the magnets you might be using to attach notes to your refrigerator are about 100 to 1000 times stronger than the Earth's field, or about 10 billion times weaker than that of a pulsar. Pulsar magnetic fields are so strong that they they would severely distort the structures of atoms exposed to them, stretching them into cigar-like shapes.

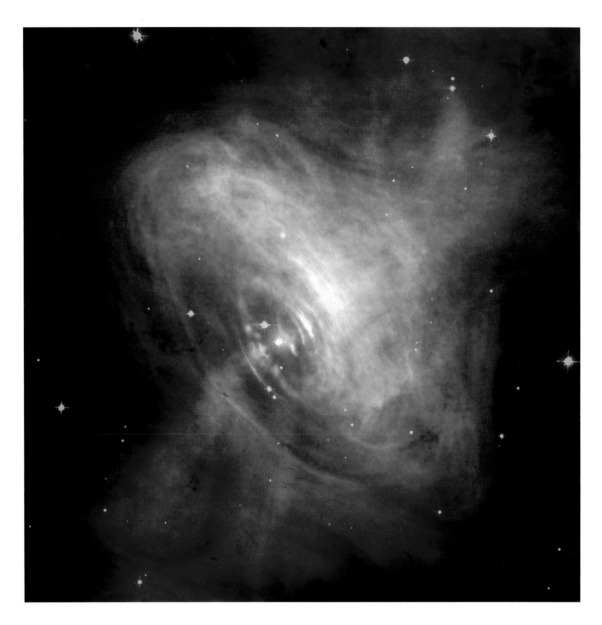

Processes occurring in the magnetic field outside a neutron star are believed to generate the "lighthouse beam" of radio waves that we see from pulsars. The field also supplies the coupling that allows the pulsar to operate as a highly efficient particle accelerator, converting its spin energy into the very energetic electrons and positrons that keep the Crab Nebula shining. The radiation intensities produced by pulsars are extreme, surpassing those in terrestrial laser beams. According to one theory, the interaction of this radiation with the intense magnetic field gives rise to the beamed emission: in such an environment, numerous "electron–positron pairs" – literally, pairs of particles consisting of an ordinary electron and its antimatter counterpart – are created *from the vacuum of space* by a process never observable in terrestrial laboratories. These particles could subsequently produce the radiation that we observe.

Neutron stars are extreme, too, in the strength of their gravitational fields: the force of gravity at the surface is 10^{12} times that on Earth. Their surfaces would be smooth: no mountain could stand more than a millimeter high. But so strong is gravity that the energy expended in climbing a millimeter mountain on a neutron star would still exceed that needed to escape entirely from the Earth's gravitational field. Because no structure would escape being squashed on the surface of a neutron star, the example of the mountain climber is even more far-fetched than it sounds. One could only consider observing a neutron star from a free-falling or orbiting platform. But even there, conditions would be uncomfortable because of the tidal effects – the difference between the gravitational pull on head and feet.

A rocket would have to attain about half the speed of light in order to escape from the surface of a neutron star, compared to about 11 kilometers per second (1/27 000 the speed of light) to escape from

This composite image of the Crab Nebula was made by superimposing data from the *Chandra X-ray Observatory* (blue) with a visible-light image from the *Hubble Space Telescope* (red). The pulsar, seen as a star-like object at the center of the innermost ring of X-rays, powers the radiation at all wavelengths. The X-rays are concentrated into a smaller region than the visible light because the X-ray emitting electrons radiate away their energy more quickly than the less energetic, optically emitting electrons, and therefore cannot travel as far from the source.

The internal structure of a neutron star is shown in cross section for each of three theoretical models. Matter near the core is more than 100 trillion times denser than water; its density exceeds even that of an atomic nucleus. The star depicted here has a mass 1.4 times that of the Sun, yet its radius is only somewhere between 10 and 16 kilometers, the size of a typical city. The models differ because they make different assumptions about the pressure of superdense nuclear matter, about which little is known.

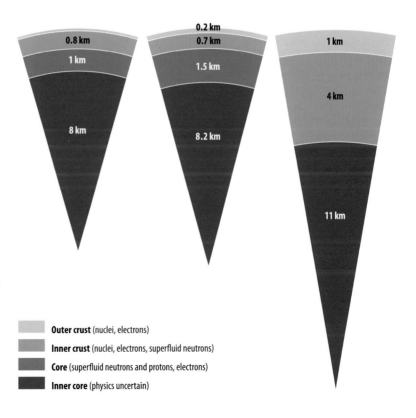

Outer crust	(nuclei, electrons)
Inner crust	(nuclei, electrons, superfluid neutrons)
Core	(superfluid neutrons and protons, electrons)
Inner core	(physics uncertain)

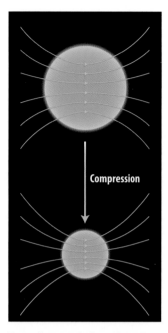

Compression

Lines of magnetic force threading an ionized gas, such as a stellar interior, follow the gas motions. When a stellar core collapses to form a neutron star, the magnetic field lines running through it are compressed. Since the density of field lines determines the strength of the field, this process can explain how neutron stars can acquire the enormous magnetic fields inferred to exist in pulsars.

the Earth. The gravitational binding energy – the energy that would have to be supplied in order to disperse the star – is 10 to 20 percent of the so-called rest-mass energy – the energy that would be released by converting all the matter in the star to energy, according to $E = mc^2$. This is a truly enormous amount of energy. The most efficient nuclear reactions produce less than 1 percent of mc^2, while chemical reactions typically convert only a few billionths of a percent. Indeed, it is from gravitation that the magnetic and rotational energies of a pulsar are derived. Just as a skater spins faster by drawing her outstretched arms to her side, an imploding stellar core increases its rotation rate, and its rotational energy, as it contracts. This faster rotation is a consequence of the conservation of angular momentum. And just as the skater must expend energy drawing her arms in (against the centrifugal force, which tries to keep them outstretched), so the gravitational force does work on the star that goes into the increased rotational energy.

In an analogous way, an ordinary stellar magnetic field can be amplified by a factor of 10 billion or more by being squeezed during the compression to a neutron star. Lines of magnetic force in an ionized gas, such as makes up the interior of a star, behave as if they are "frozen-in," meaning that they are dragged along wherever the gas goes. Since the strength of a magnetic field is determined by the density of field lines, the magnetic field in a collapsing stellar core can increase by a huge factor. Here, the work done by gravity against the "magnetic pressure" force is what builds up the magnetic energy. Thus, gravitation not only energizes the supernova explosion itself but also leaves a "dead" star with more energy than nuclear reactions could generate over its entire previous lifetime. Gravity is the ultimate source of the energy that powers the Crab Nebula.

Before leaving the Crab Nebula, we offer an historical speculation. The nebula's central pulsar can be seen through any large optical telescope, but the pulse repetition rate is so high that the eye responds to it as a steady source. But had the pulsar been spinning, say, 10 times a second rather than 30 times, the remarkable properties of the little star in the Crab Nebula could have been discovered eighty years ago, long before the development of radio astronomy. How would the course of twentieth-century physics have been changed if superdense matter had been detected in the 1920s, before neutrons were discovered on Earth? One cannot guess, except that astronomy's importance for fundamental physics would surely have been recognized far sooner.

Black Hole Stellar Remnants

Gravitational collapse does not appear to be a fate that stars work actively to avoid. Stars can shed a great deal of mass during their

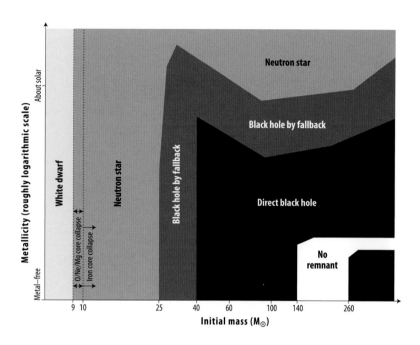

This diagram (adapted from Alexander Heger *et al.*) illustrates the probable fates of stars with different initial masses (horizontal scale) and chemical compositions (as measured by the concentration of elements heavier than helium, or "metallicity": vertical scale). Stars of high metallicity are thought to lose much of their mass by stellar winds – this is why even very massive stars of high metallicity are believed to form neutron stars. In the region labeled "black hole by fallback," the supernova explosion initially produces a neutron star, but enough matter subsequently "falls back" onto it to cause it to collapse. Stars in the dark region at lower right are expected to collapse directly to a black hole, except in the white band where a "pair-instability" supernova explosion might disperse the star completely, leaving no remnant. Rotation, which is also expected to affect the outcome of stellar collapse, is not taken into account in this diagram.

evolution – those of up to 5 (or even 8) solar masses can end up as white dwarfs. But the discovery of neutron stars in supernova remnants proves that still heavier stars do not shed enough gas to bring themselves safely below the Chandrasekhar limit. The upper mass limit for a neutron star is 2 or 3 solar masses, not very much larger than the limit for a white dwarf. No theoretical arguments suggest that stellar evolution should permit the existence of remnants too massive to become white dwarfs while forbidding those too massive to become neutron stars.

Stellar remnants more massive than about 2 or 3 solar masses therefore seem fated to undergo complete gravitational collapse to black holes, because no form of internal pressure can maintain them in equilibrium once their nuclear energy sources are exhausted. These gravitationally collapsed bodies may not be quite so common as neutron stars, but still there may be millions of them in our Galaxy. Some types of supernovae may form black holes rather than neutron stars; or, some massive stars may collapse without producing any conspicuous supernova-type explosion.

Where are these alleged black hole remnants? Can we identify them? Theoretical arguments are nourishing up to a point, but we must demand empirical evidence before we can replace conjecture with confidence. Fortunately, the development of a new kind of observational astronomy in the 1960s – X-ray astronomy – led straight to some very convincing evidence for stellar-mass black holes in our own Galactic neighborhood.

3

BLACK HOLES IN OUR BACKYARD

How would you go about searching for the black hole remnant of a massive star in our Galaxy? When this first became a serious question among astronomers early in the 1960s, surprisingly few of the experts on black holes bothered to ask it. The reasons, in part, were "cultural." C. P. Snow had written about "two cultures" – scientific and humanistic – diverging in language and habits of thought, and becoming less and less able to communicate with each another. But cultural divergences had occurred *within* the physical sciences as well. Black holes, with their elegant mathematical properties, were primarily the intellectual toys of a highly specialized group of relativists – that is, scientists specializing in the subtleties of general relativity. In the tradition of Einstein, relativists liked to rely on pure thought and mathematical elegance to deduce deep truths about the nature of the Universe. On the other hand, astrophysics had a strong empirical tradition. Theory was important, but it was (and remains) observation that drove progress in the field.

The discovery of black holes in nature required synergism between astronomers and relativists; fortunately, just such an alliance had developed by the mid-1960s. Perhaps the "send-off" was the First Texas Symposium on Relativistic Astrophysics, which was held in Dallas in December 1963 in the aftermath of the discovery of the extraordinarily luminous and distant objects known as quasars. There the astrophysicist Thomas Gold (who in 1968 became the first to identify pulsars as rotating magnetized neutron stars) heralded the collaboration in his famous after-dinner speech, suggesting "that the relativists with their sophisticated work are not only magnificent cultural ornaments but might actually be useful to science! Everyone is pleased: the relativists who feel that they are being appreciated and are experts in a field [i.e., relativistic astrophysics] they hardly knew existed; the astrophysicists for having enlarged their domain, their empire, by the annexation of another subject – general relativity. It is all very pleasing, so let us hope that it is right. What a shame it would be if we had to go and dismiss all the relativists again."

Perhaps, too, the observational prospects for detecting black holes had been largely ignored because success seemed unlikely. But the possibility of failure did not deter the Soviet theorist Yakov Zel'dovich, the American Edwin Salpeter, and a few others from boldly speculating on possible observational signatures. Once formed, black holes are essentially "passive," and the best hope of locating them lies in discerning their gravitational effects on neighboring stars or gas. Zel'dovich's first proposal was to search for binary star systems in which an ordinary star is orbiting an invisible, yet massive, companion. By applying Newton's laws (which would still hold quite accurately far outside the event horizon of the black hole) to the measured speed and orbital period of the normal star, one can sometimes estimate the mass of the unseen companion.

By the 1960s this technique had already been long established as the most direct method for estimating stellar masses. However, it is usually fraught with uncertainties, and does not yield unique values for the masses of both stars. At best one could hope to obtain a lower limit on the mass of the unseen star. While the method could work in principle, to this day Zel'dovich's initial suggestion of searching for an invisible star has not yielded any plausible black hole candidates.

But Zel'dovich and Salpeter also came up with other ideas. They noted that a black hole moving through space would absorb matter from the tenuous interstellar medium – an atmosphere of gas and dust that fills the volume between the stars in our Galaxy. This taking in of matter is the process called accretion, of which we shall hear much more later in this chapter. What Zel'dovich and Salpeter realized was that a black hole accreting interstellar gas might conceivably be quite luminous. As the gas falls deep into the gravitational field of the black hole (but before it crosses the horizon, of course), it will be strongly compressed and heated. In principle, the hot infalling gas could radiate into space as much as a few percent of the energy locked up in its mass, according to Einstein's relation $E = mc^2$. If reality matched theory, a black hole accreting a given amount of matter could produce more than 10 times as much energy as the most efficient thermonuclear incineration of the same material.

In a synthesis of these two proposals, Zel'dovich and his former student Igor Novikov proposed in 1966 that the best place to look for compact stellar remnants – either neutron stars or black holes – was in close binary systems, where the black hole or neutron star could capture and accrete matter from its stellar companion. Shock waves in the falling gas would heat the gas to temperatures on the order of 10^8 degrees Kelvin; at these temperatures the gas would emit radiation predominantly in the X-ray band. Rapid, random flickering of the X-rays, on timescales of a few milliseconds, was later recognized as an additional indication that the emission was being produced in an extremely compact region with the dimensions of a black hole or neutron star.

These signatures are exactly what led astronomers in the 1970s to the discovery of stellar-mass black holes, along with close binaries in which the compact accreting object is a neutron star rather than a black hole. Insightful as it proved to be, this lucky guess was not taken in an observational vacuum. Even the most brilliant theoretical astronomers have seldom foreseen the startling variety of real phenomena that appear in the sky. At the time that Zel'dovich and Novikov made their suggestion, the journals and conference circuits were already abuzz with the amazing discoveries being made using an entirely new kind of observatory, one that gave us a view of the Universe through "X-ray eyes." It

Yakov Zel'dovich (right) and Igor Novikov. After a versatile earlier career, Zel'dovich became an outstanding researcher in relativity and cosmology. With Igor Novikov and other younger colleagues in Moscow, he was, from 1960 onward, responsible for many of the key ideas in relativistic astrophysics. Zel'dovich died in 1987; Novikov later headed a research group in Copenhagen, Denmark.

is hard to know the extent to which their suggestion was stimulated by the debate then going on over the nature of the first X-ray "star" to be discovered, Scorpius X-1.

X-Rays from the Cosmos

Physicists at the turn of the twentieth century knew well that heated gases give off radiation over a wide range of wavelengths, and that extremely hot gases emit X-ray and even gamma-ray radiation. But until the middle of the century, astronomers' view of the sky was restricted to the narrow band of wavelengths centered on visible light. If we had to identify one stimulus for the revolution in our conception of the Universe during the past sixty years, it would have to be the opening up of the electromagnetic spectrum to observational astronomy, first in the radio band (as we will discuss in Chapter 6), then in the X-rays, and more recently, in the infrared, ultraviolet, and gamma-ray bands. Rather than just confirming long-held theories, each advance into a new wavelength regime revealed entirely unexpected phenomena.

When we use an X-ray telescope to pinpoint and measure the sources of X-rays on the sky, we now know that we are studying the most energetic features of the Universe – the hottest gases, the most intense gravitational fields, the most energetic explosions. X-ray observations provided the first strong evidence that black holes actually exist. But observing the sky at X-ray wavelengths was far from easy. Unlike visible light, X-rays cannot penetrate very far into the Earth's atmosphere. To observe them, we must send detectors above the atmosphere, using rockets and satellites. X-ray astronomy was truly the first "new astronomy" of the space age.

Our first glimpses of the X-ray sky came from detectors carried on board sounding rockets, small boosters originally designed to make measurements of the atmosphere. Each suborbital flight yielded only about 4 minutes of useful data, enough to detect X-rays from the Sun, as expected. But in 1962, a rocket flown by the American Science and Engineering (AS&E) group, led by Riccardo Giacconi, discovered an apparently discrete source of X-rays from beyond the Solar System in the direction of the constellation Scorpius. This became known as the source Scorpius (or "Sco") X-1. Now we know that Sco X-1 contains a neutron star orbiting a normal star, and that the X-rays are produced close to the surface of the neutron star. A subsequent rocket flight by Herbert Friedman's group at the U.S. Naval Research Laboratory (NRL), using improved apparatus, provided a more accurate location for Sco X-1 and also detected X-rays from the direction of the Crab Nebula, the remnant of Yang Wei-Te's supernova of A.D. 1054.

In the early days of space astronomy, observations were made from "sounding rockets," which gathered only a few minutes' worth of useful data. These pictures show the launch and "recovery" of an experiment to study the cosmic X-ray background – a diffuse X-ray glow coming from all points in the sky. The experimenter, Andrew Fabian, obtained his Ph.D. on the basis of this work, and has gone on to become a leading X-ray astronomer.

The discovery of Sco X-1 was a great surprise. Astronomers had been pessimistic about the prospect of detecting X-rays from beyond the Solar System, since the X-ray emission from known stellar objects was expected to be too weak to be picked up with any feasible instruments. Sco X-1, emitting thousands of times more energy in the X-rays than in visible light, was clearly an entirely new and unsuspected type of source. The same AS&E rocket flight also revealed an unexpectedly strong X-ray glow in all directions in the sky, now known as the *X-ray background*. The discoveries from these experiments – each based on just a few minutes of data – provided the initial impetus to the entire subject of X-ray astronomy.

X-ray astronomy took a major leap forward when detectors were flown on satellites. Now it was possible to observe faint sources for extended periods of time and to make surveys of the whole sky, neither of which was feasible during brief rocket flights. The first such payload was conceived, built, and operated by Giacconi and his colleagues at AS&E. It flew on NASA's small *Uhuru* X-ray satellite, launched in 1970. (Launched from a platform off the coast of Kenya on the seventh anniversary of that country's independence, the satellite was named with the Swahili word

Riccardo Giacconi's experiments using sounding rockets led to the discovery of the first cosmic X-ray source (Sco X-1) and the cosmic X-ray background. He directed the pioneering *Uhuru* satellite project, followed by the much more ambitious *Einstein Observatory*. Giacconi was the first Director of NASA's Space Telescope Science Institute and later served as Director-General of the European Southern Observatory, where he oversaw the construction of the world's largest optical telescope. He was awarded the 2002 Nobel Prize in Physics for the discovery of cosmic X-ray sources.

for "freedom.") *Uhuru* detected numerous extragalactic X-ray sources, including "active" galaxies emitting intense concentrations of energy from their centers and clusters of galaxies. But *Uhuru's* remarkable success stemmed from its unexpected discovery of a whole new class of sources, the binary X-ray sources, of which Sco X-1 is a member.

X-ray binaries have a wide range of properties. They include sources like Hercules X-1 and Centaurus X-3, which pulsate regularly every few seconds, and sources like Cygnus X-1 – the first black hole candidate – which seem to flicker at random on timescales down to a few milliseconds. Other sources occasionally undergo large outbursts that correspond to the ejection of powerful winds or jets. These are called "X-ray novae" or "microquasars," by analogy with much more powerful objects we will discuss in Chapter 5. What makes these sources behave differently from one another? For reasons that will become clear shortly, Cygnus X-1 and most of the microquasars have been identified as black holes. On the other hand, the regularly pulsing sources in binary systems are interpreted as spinning neutron stars, sometimes called *X-ray pulsars*. The masses of about a dozen of these periodic compact sources have been determined from the Newtonian dynamics of the binary system. All of the masses lie between 1 and 2.5 times the mass of the Sun, below the likely mass limit for neutron stars.

Although these neutron stars pulsate with a steady frequency at X-ray wavelengths, the radiation mechanism is apparently very different from that of *radio pulsars*, like the one in the Crab Nebula. The intense X-rays are generated by infalling gas, captured from the companion star, impacting on the neutron star's surface at one-third to one-half the speed of light. The kinetic energy of the gas is converted to thermal energy on impact, and radiated at temperatures of roughly 10 to 100 million degrees Kelvin, giving predominantly X-rays. The explanation for the regular pulsations is that the gas does not fall uniformly over the surface of the neutron star, but is channeled onto the magnetic polar caps by the extremely strong field. As the neutron star spins, the X-rays from the pair of "hotspots" come in and out of view, producing regular pulses. Some radio pulsars, including the Crab, also pulse in X-rays, but with not nearly the same intensity.

It was an accident of timing that neutron stars were discovered by radio astronomers (in their manifestation as pulsars) rather than by X-ray astronomers. The X-ray astronomers had been expecting to discover neutron stars, because at the time people thought that the surfaces of neutron stars might be very hot. (Isolated, nonpulsing neutron stars have now been detected, and they are much cooler than had been expected.) In the earliest days of rocket experiments, Friedman's group at NRL observed the Crab Nebula during the few minutes it was covered on the sky by the Moon. Since the position of the Moon's limb was known very accurately, this lunar occultation experiment was a unique opportunity to probe the structure of

This diagram shows the masses of compact objects that are believed to be either neutron stars or black holes. The best determinations are of binaries containing two neutron stars. Other neutron star masses are not so precisely determined (as indicated by horizontal error bars), but all lie in the range consistent with their being spinning neutron stars (to the left of the dashed vertical line). In contrast, the black hole candidates shown in the diagram have much higher masses, with error bars that do not extend down into the neutron star range.

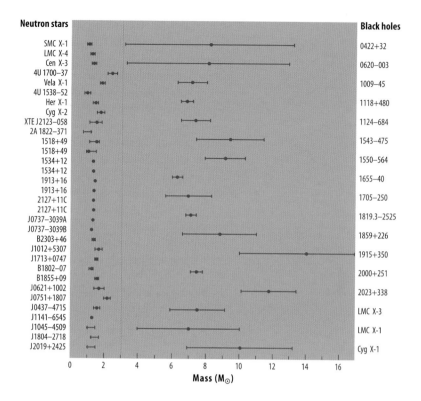

In an X-ray pulsar, the neutron star's enormous magnetic field diverts the flow of infalling gas and channels it onto the magnetic polar caps. The impact of the gas on the star's surface creates a pair of X-ray hotspots. The star's rotation produces the appearance of periodic X-ray pulsations.

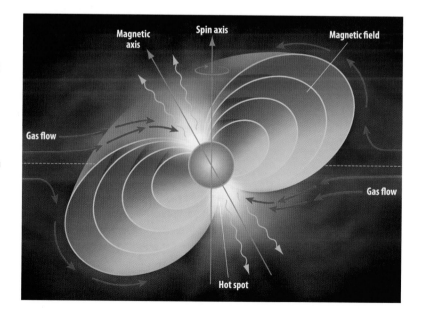

the Crab X-ray source with unprecedented angular resolution. The purpose was to test whether the X-rays came from a point source such as a hot neutron star or from the extended nebula around it. The result – that the Crab X-ray source was extended – was corroborated by later work. But had Friedman's experiment been slightly more sensitive, he would have found that 10 percent of the Crab's X-rays do indeed come from a central point source, and thus he would have discovered the X-ray emission of the Crab pulsar. He did not have an accurate clock in his detector, however, and would not have discovered that the source was flashing 30 times every second. But the X-rays certainly would have tipped off investigators that this was a likely detection of a neutron star. One wonders how much longer it would have taken before the pulsations were discovered. (Data recorded during a 1964 balloon experiment by Richard Haymes of Rice University actually contained evidence of regular pulsations in the X-rays from the Crab, but this was only discovered when the data were reanalyzed several years later.)

On the other hand, had Jocelyn Bell been less perceptive in 1967, neutron stars might have escaped notice by radio astronomers, and their discovery would have been deferred until the launch of the *Uhuru* satellite, three years later. The X-ray sources Her X-1 and Cen X-3, discovered by *Uhuru* in 1971, would have been quickly interpreted as neutron stars. It was thus bad luck for the X-ray researchers that radio astronomers won the credit for discovering neutron stars – X-ray astronomers had been seeking them actively in the 1960s, and did in fact discover them by a quite different route three years later. However, even though X-ray astronomers lost the glory of discovering neutron stars, X-ray data have subsequently yielded important information about neutron star physics, complementary to what can be learned from radio observations.

Whereas regular pulsations are a sure giveaway that an X-ray-emitting object is a neutron star, such pulsations are not always present. Many sources, like Sco X-1, are suspected of harboring neutron stars even though they show no signs of regular pulsations. Why, then, are we convinced that the X-ray source in Cygnus X-1 is a black hole and not a neutron star? Distinguishing a black hole from a neutron star in such a system is not trivial, since the gravitational field near a neutron star is nearly as strong as that near a black hole. The effects on infalling gas are broadly similar; therefore, any differences in the radiation emitted in the two cases are likely to be subtle. There are probably some differences in the spectrum of the radiation as well as the nature and strength of the random flickering. After all, neutron stars have real surfaces whereas black holes have horizons, which represent the complete absence of a physical surface. The magnetic fields and rotation (however slow, or masked by an opaque screen of gas) of neutron stars would also imprint features on the radiation that would be absent in a black hole. But no one has succeeded in predicting the differences purely from theoretical

Focus

Hyperbolic surfaces

Parabolic surfaces

X-rays   **X-rays**

X-rays cannot be focused by conventional mirror arrangements like those used in optical telescopes. In X-ray telescopes like the ones flown on the *Einstein* satellite and *Chandra X-ray Observatory*, X-rays are focused into an image by reflecting off the insides of tapered metallic tubes at very shallow angles, or "grazing incidence." To achieve a large total "collecting area" and therefore high sensitivity, many of these reflecting surfaces are nested inside one another. This mirror system gathers the X-rays entering the telescope through an annulus as shown.

considerations. In fact, having used other methods to select black hole "candidates," astronomers have been trying to determine the spectral signatures of black holes empirically, with mixed success so far.

Fortunately, there is a decisive discriminant that can at least rule out the presence of a neutron star from some X-ray binary systems. If one can show that the compact object has a mass greater than 3 solar masses, then one can exclude a neutron star. In lieu of any other plausible candidates for a massive compact object capable of producing flickering X-rays, these systems are thought to contain black hole remnants of stars. About 20 X-ray binaries are now known to meet this criterion.

The spectacular and highly publicized early successes of X-ray astronomy led to a rapid development of the field. Following *Uhuru*, several small X-ray satellites were launched during the 1970s, signaling the entry of other nations into a field that had hitherto been predominantly a U.S. enterprise. These projects consolidated the discoveries made with *Uhuru*, and discovered additional kinds of X-ray binaries. The American *Einstein* satellite, launched in 1978, carried the first X-ray telescope combining high angular resolution with a large collecting area – an innovation, made possible by the telescope's nested mirror configuration, that offered much higher sensitivity. The results from *Einstein* had broad ramifications for many areas of astronomy, including the search for black holes in both X-ray binaries and the nuclei of galaxies. Major projects in X-ray astronomy have by now been carried out by all spacefaring nations, the latest being the American *Chandra X-ray Observatory*, the European *XMM–Newton* telescope, and the Japanese *Suzaku* satellite. The progress that has led from simple detectors in sounding rockets to major telescopes that offer images as sharp as those made by big optical telescopes, spectroscopic capability, and phenomenal sensitivity constitutes a prime scientific success of the space program.

Mass Transfer and Accretion in Binaries

In all X-ray binaries, matter is being transferred from a normal star to a compact object, whether neutron star or black hole. The infall of gas onto a star or other gravitating system is called *accretion*. Astronomers first considered the possibility of a star's accreting matter when they realized that the space between the stars is not a perfect vacuum but is pervaded by a tenuous gas (the interstellar medium).

Astronomers' early interest in the process had nothing to do with the emission of X-rays (or any other form of energy) by a star, but rather, with examining whether a star could grow appreciably in mass by accretion, or at least undergo an alteration of the chemical composition of its surface. (The latter is still thought to be a viable mechanism for explaining certain chemical

This diagram depicts how accretion can occur onto a black hole moving through diffuse gas. In the hole's frame of reference, the gas is flowing past as though in a wind. The hole's gravity causes the flow to converge into a "wake," and a shock front forms "downstream." The shocked gas can lose energy and fall into the black hole. This process was first discussed (with application to normal stars rather than to black holes) by Hermann Bondi, Fred Hoyle, and Raymond Lyttleton.

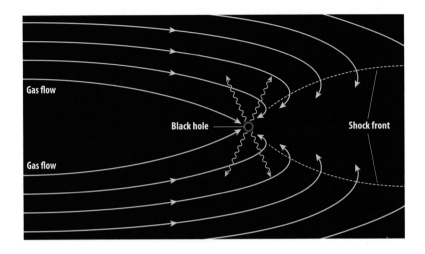

abundance anomalies in white dwarfs.) In 1926, Eddington considered the rate at which a star would sweep up matter, taking into account the enhancement due to an effect called gravitational "focusing" (not to be confused with gravitational "lensing," which we discuss in Chapter 4). The rate at which gas particles would hit a nongravitating sphere is simply determined by the surface area of the sphere and the speed and density of the particles. But gravity has the effect of focusing the trajectories of the gas particles toward the center of mass, so that some particles that would have missed the nongravitating sphere will hit the gravitating one. The accretion rate would therefore be larger by a factor that depends on the mass. The Cambridge astrophysicists Fred Hoyle and Raymond Lyttleton took Eddington's argument further by showing that the accretion rate would be greatly enhanced if particles were colliding close to the star. This idea was extended and formalized by Hermann Bondi in 1952, and most astronomers believe that the Bondi–Hoyle–Lyttleton theory gives a fair picture of accretion by a star moving through a gaseous background.

The idea that mass can be transferred between stars in a binary system also predates the discovery of binary X-ray sources. The star Algol (from *al ghoul*, Arabic for "the demon") in the constellation Perseus is an eclipsing binary with a special place in the history of stellar evolution studies. Every 2.9 days the smaller (but brighter) blue star is hidden from view by a larger (but dimmer) red star. The brightness of the binary source drops by a factor of 3 and the apparent color changes. The orbital parameters are well enough known that the masses of both stars can be determined, leading to the famous "Algol paradox." It turns out that the blue star, which is still burning hydrogen in its core in its main sequence phase, is more massive than the red star, which is already on its way to becoming a red giant. But stellar evolution theory says that the more massive star should become a red giant first, if we assume (as must surely be the case) that both stars formed at the same time.

The evolutionary history of the Algol binary explains the "Algol paradox." As the initially more massive star swelled into a red giant, it overflowed its Roche lobe and dumped so much matter onto its main sequence companion that the relative masses were reversed. According to stellar evolution theory, if a binary contains a red giant and a normal star, the red giant should be the more massive star. Because of mass transfer, this rule is violated in Algol – hence the "paradox." The light curve plotted shows the periodic eclipses, which tell us not only the orbital period but also the relative sizes and brightnesses of the two stars; spectra reveal Doppler shifts that tell us how fast the stars are orbiting about each other; with this information we can determine the masses. Astronomers can therefore learn a great deal about Algol (and similar systems) even though the stars are too close for optical telescopes to separate the two images.

The resolution of the paradox is that the less massive star *used to be* the more massive star and did indeed reach the red giant phase first. As it swelled, however, it overflowed its own zone of gravitational influence (its teardrop-shaped *Roche lobe*) and spilled across the *Lagrange point* into its companion's zone of influence. The companion absorbed so much mass in this way that it became the more massive star.

If we follow this process to its logical conclusion in a sufficiently massive binary system, we can understand how an X-ray binary can form. The details were first worked out by the Dutch astrophysicist Edward van den Heuvel in the early 1970s, and they have been refined by many workers since. We start out with an Algol-type situation in which mass transfer has left a massive star heavier than 5 solar masses that has evolved beyond the main sequence and an even more massive star still on the main sequence. Although the evolved star has lost most of its mass, it proceeds steadily on its march toward gravitational Armageddon (since its rate of evolution is

Evolution of a close binary system into an X-ray binary, with cumulative times shown. In this example, the two stars are initially 20 and 8 solar masses. The heavier star is the first to exhaust its hydrogen fuel. It swells up into a giant and loses mass, most of which is dumped on its companion. Its remaining core, however, continues to evolve and explodes as a supernova. Because the exploding star is by this stage less massive than its companion, the sudden mass loss is not enough to disrupt the binary. The compact remnant stays in orbit around its companion – a (now much more massive) bright blue star with a strong stellar wind. Accretion of gas from this wind gives rise to a strong compact X-ray source.

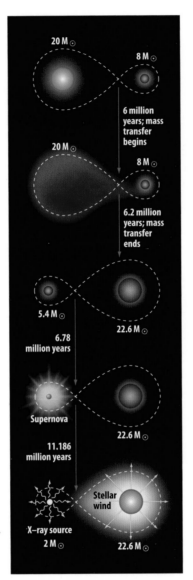

determined by the *core* of the star, which does not participate in the mass transfer), and within a few hundred thousand years it blows up as a supernova. Here is where a crucial aspect of the Algol-type configuration comes in. If the *more* massive star in a binary explodes, leaving behind a neutron star or black hole, the result is almost always to destroy the binary: the compact star and its former companion go flying off in different directions. But if the exploding star is the *less* massive member of the system, the binary remains intact. The latter condition is automatically satisfied in an Algol-type binary. So we have a binary consisting of a neutron star (or black hole) and a very massive main sequence star. Now it is the latter that transfers mass onto its now-compact companion. This mass comes at first from a strong stellar wind; within a few million years the massive star evolves into a red giant and overflows its Roche lobe.

A second phase of mass transfer, onto the neutron star or black hole, thus follows the first phase, which set up the Algol-type configuration. This second phase of mass transfer is more dramatic than the first, for a very simple reason: the gravitational field near the surface of the neutron star (or the horizon of the black hole) is much stronger than the gravitational field near the surface of any ordinary star. When matter is accreted by a gravitating body, energy is released in what we can conceptualize as a two-stage process. (In reality, the two stages occur simultaneously.) First, gravity accelerates the gas and imparts kinetic energy to it. Since energy can neither be created nor destroyed, but can only change from one form to another (that is, energy is *conserved*), this kinetic energy must come from somewhere. Its source is the gravitational interaction between the gas and the compact star. We say that matter in the presence of a gravitational field possesses gravitational *potential* energy, which is converted to energy of motion through gravitational acceleration. In the second stage of the energy-release process, the kinetic energy is *dissipated*, being ultimately converted to heat or radiation or both. The exact way in which this energy conversion occurs can be complicated, and may involve collisions among turbulent cells of gas, interactions with a magnetic field, and other processes. However, the "bottom line" – the net amount of energy available to be radiated – is determined by the conservation of energy and is insensitive to all the messy details: it just depends on the amount of gravitational potential energy released as the material drops deeper and deeper into the gravitational field.

We can quantify the amount of energy released per gram of matter falling into a gravitational field by specifying the depth of the *gravitational potential well*. This quantity is often expressed in units of mc^2, the energy equivalent of the mass being dropped into the field. As a useful reference, thermonuclear burning of some amount of material would release at most 0.8 percent of mc^2. Expressed in units of mc^2, the gravitational field of the Sun is puny. The gravitational potential at the surface is only a few times 10^{-6}, meaning that you would get several thousand times more energy out of a gram of matter by placing it in the core of the Sun and burning it than by dropping it onto the surface. For neutron stars and black holes, the potential is about 0.1, or more than 10 times the peak efficiency of nuclear burning.

As sources of accretion power, white dwarfs are intermediate between ordinary stars and neutron stars or black holes. The gravitational potential of a white dwarf is about 100 times higher than that of the Sun, but is still considerably lower than the maximum energy released by nuclear burning. Nevertheless, binaries containing accreting white dwarfs produce spectacular effects, through both gravitational energy release and nuclear reactions. A white dwarf by itself will not support nuclear reactions, but such reactions can sometimes take place in a layer of "fresh," hydrogen-rich nuclear "fuel" deposited on the surface by accretion. Since white dwarfs are so faint to begin with, the release of gravitational potential energy alone can create an interesting observational show, characterized by intense and highly variable flares of ultraviolet radiation. Accreting white dwarfs in binaries are generally known as "cataclysmic variables." They include "dwarf novae," and recurrent "novae," as well as other variants, depending on the nature of the accretion and whether the white dwarf has a strong magnetic field.

Generally, the gas settling onto a white dwarf is not hot enough to support steady thermonuclear burning, but as more material accumulates the gas near the bottom of the layer is compressed by the weight of the overlying gas. If the density becomes large enough it can trigger *pycnonuclear* reactions (from the Greek *pyknos*, for "dense"), which, in contrast to *thermonuclear* reactions, do not require a high temperature. The high density alone causes the atomic nuclei to develop high random speeds, according to the Pauli exclusion principle, just like the electrons holding up the interior of a white dwarf. These nuclei slam into each other with high enough speed to undergo nuclear reactions. This is a form of "cold fusion" that works, but only at densities thousands of times higher than those attained in the notorious laboratory experiments of a few years ago. Once these pycnonuclear reactions get underway, they rapidly heat the gas to temperatures high enough for thermonuclear reactions to take over, and the energy generation rate climbs steeply. The nuclear energy released easily overwhelms the gravitational binding energy of the accreted layer, and the layer is blown off the star in an explosive event called a *nova*. Unlike a supernova, which permanently alters or destroys the exploding star, a nova is only a surface event, and can repeat every time a new layer of gas is accreted.

What happens if a white dwarf accretes so much matter that it slightly exceeds the Chandrasekhar limit of 1.4 solar masses? It will start to implode, and if no nuclear reactions take place it will certainly collapse all the way to a neutron star. But if there is substantial nuclear fuel left to burn inside the white dwarf, then the collapse itself may trigger the onset of runaway nuclear burning. Provided that nuclear burning begins before the collapse has proceeded too far, the energy released by the nuclear reactions will be sufficient not only to reverse the collapse but to completely disperse the star. The energy released in this thermonuclear explosion, which is called a "Type Ia" supernova, is very similar to the kinetic energy carried off by the remnants of the gravity-powered core-collapse supernova (usually labeled Type Ib or Ic, or Type II) we discussed in the last chapter. But the total energy released is far greater in a core-collapse supernova. The reason they appear to be comparable is that 99 percent of the energy released in a core-collapse supernova is carried away by nearly undetectable neutrinos. Only 1 percent is released as kinetic energy, whereas a Type Ia supernova puts most of its energy into the expansion of its ejecta. So the gravity-powered explosion wins energetically, even if the competition appears to be a draw!

Neutron stars and black holes are the real gravitational powerhouses; their gravitational potentials of roughly 0.1 mean that infalling objects will release an amount of energy equivalent to 10 percent of their mass. There is no way that steady nuclear burning can compete with this rate of energy release, due entirely to gravitational infall. Occasionally the accumulated nuclear fuel on the surfaces of certain accreting neutron stars flares up, producing *X-ray bursts*. Unlike a nova explosion on a white dwarf, these bursts are not able to blow the surface layer off the neutron star, but they do dominate the star's luminosity for a brief time. Over the long haul, however, it is accretion power that reveals both neutron stars and black holes in close binary systems, so it is essential that we understand the accretion process as thoroughly as possible.

Accretion Disks

The accretion of matter from a binary companion is a much more complex process than accretion from a uniform gaseous background. Since the stars in the binary are orbiting each other, any gas flowing from one star to the other does so with a lot of sideways motion or, more precisely, angular momentum. It cannot simply fall "straight in." Like water going down a drain (which generally has too much angular momentum to go straight in), the gas forms a kind of vortex centered around the accreting object.

Gas captured by a neutron star or black hole would generally not fall radially inward: because of its angular momentum, it forms a swirling disk or vortex. Arthur Rackham's illustration for Edgar Allen Poe's *A Descent into the Maelström* dramatizes the effect of a whirlpool.

There are some important differences between a vortex in a bathtub and the gas swirling around a black hole or neutron star. Gas is always much more compressible than water, and it becomes even more so as it cools down. The accreting gas in most X-ray binaries cools quite effectively, so instead of a three-dimensional vortex the flow gets flattened down to a spinning platter-shaped configuration, called an *accretion disk*. (There are some conditions under which an accretion flow may not flatten into a thin disk, especially flows into supermassive black holes in "quiet" galactic nuclei, which we will discuss in Chapter 8. In X-ray binaries this is probably more the exception than the rule.) The platter in this case does not rotate rigidly, however, in contrast to, say, a spinning phonograph record or compact disc. Instead, the inner parts are swirling around more rapidly than the outer parts. It is this *differential rotation* that allows the star or black hole to absorb the gas, since the rapidly rotating inner parts drag against the slower outer parts, lose some angular momentum, and spiral slowly inward. Basically the same effect allows water to go down the bathtub drain rather than remain in the vortex indefinitely.

For inflow to occur, there must be some kind of *friction* between the more rapidly spinning parts of the disk and the slower parts. In the bathtub, we know what causes this friction: the viscosity of the water, arising from interactions among water molecules. But under astrophysical conditions, in an accretion disk or almost anywhere else, this kind of "molecular" viscosity is much too small to be important. Uncertainty about the nature and magnitude of the "viscosity" has given accretion disk theorists a chronic headache from the earliest days of the subject. A promising possibility is that irregular, rapidly changing magnetic fields give rise to a kind of viscosity. Calculations show that these magnetic field fluctuations can arise spontaneously, as a consequence of the disk's differential rotation, but the strength of the viscosity effect remains uncertain. Fortunately, many of the observable predictions of accretion disk theories, such as the spectrum of the radiation and characteristic timescales of variability, do not depend on the exact prescription for the viscosity. A great deal of progress has been made by expressing the strength of the viscosity in terms of parameters "to be determined."

In an accretion disk, the gas closer to the center is orbiting faster than the gas farther out (higher speeds depicted by longer arrows). Frictional effects tend to transfer angular momentum (and energy) outward, allowing the gas to drift slowly inward. The nature of the friction is uncertain: it could arise from turbulent magnetic fields. The friction heats the disk, and the inner parts radiate more intensely than the outer parts. This radiant energy comes ultimately from the energy released as gas swirls deeper into the gravitational potential well of the central object.

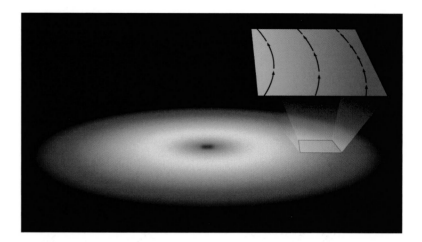

The strong magnetic field of a neutron star (and some white dwarfs) introduces further complications into the basic disk theory. If the field is strong enough, it will disrupt the disk far from the stellar surface, deflecting the accretion flow toward the magnetic polar caps. If the neutron star is spinning, the result can be the pulsations mentioned earlier, in sources like Her X-1 and Cen X-3. But if the neutron star is spinning fast enough, the magnetic field can inhibit accretion altogether, by flinging the gas away from the star.

By observing cataclysmic variables, astronomers have been able to test some key theoretical predictions for accretion disks around white dwarfs. A byproduct of frictional drag is the heating of the gas in the disk, much as the friction of a rope sliding through your hands can give you a "rope burn." According to theory, some heating should occur at all radii in the disk, and not just where the gas hits the central object at the center. The entire disk should glow, with a temperature that decreases with increasing radius. The accretion disks around white dwarfs are far too small to image directly, but Keith Horne and collaborators have produced images of accretion disks using an ingenious technique akin to medical tomography (which relies on a series of two-dimensional "slices" to reconstruct a three-dimensional image of the internal organs). The results of Horne's "eclipse mapping" and other diagnostic techniques have verified the general outline of accretion disk theory, but also present some puzzles. For example, the outer parts of disks are often hotter than predicted, even taking into account the irradiation of the outer parts by the inner parts. Also, eclipse mapping of edge-on disks show that they are often thicker than expected. These anomalies probably reflect our limited understanding of the magnetic phenomena that govern accretion; their resolution may have to wait for more sophisticated computer modeling.

In this computer simulation by Omer Blaes, Shigenobu Hirose, and Julian Krolik, magnetic lines of force are depicted as strands of "spaghetti" and are color-coded according to the deviation of gas density from its average value. This illustration represents a vertical "core" through a disk, with the red strands close to the disk equator and the green and blue strands above and below it. The chaotic structure in the red region suggests that magnetic "viscosity" is operating most vigorously there.

Stellar-Mass Black Holes: Firming Up the Case

If the credit for discovering neutron stars must be shared between the radio and X-ray astronomers, it is the X-ray astronomers who have led in the search for stellar-mass black holes in our Galactic backyard. By

Some close binaries, like UU Aquarii, are so oriented that the normal star passes in front of the accretion disk once per orbit. By observing the changes in brightness and color as the eclipse proceeds, the appearance of the disk can be reconstructed, even though it is too small to resolve with any telescope. Keith Horne has pioneered this technique.

the late 1960s, several theorists had already recognized that black holes in close binary systems should be accreting material from their companions and should manifest themselves as X-ray sources. They pointed out that such X-rays, though not varying periodically, might display irregular flickering on timescales as short as a millisecond (this being the characteristic period of the innermost orbits around a hole of a few solar masses, where most of the energy is released).

But flickering is not a guarantee that the compact object is a black hole rather than a neutron star. Indeed, Cygnus X-1 has proven to be almost alone among black hole candidates in flickering on millisecond timescales, whereas nonpulsating neutron stars often flicker vigorously. As Zel'dovich pointed out, the clincher would be if the mass of the compact object exceeded the highest possible mass of a neutron star. Such mass estimates can be obtained by measuring the orbital period and speed of the normal star using the Doppler effect. In this way Cygnus X-1, an X-ray source in a binary system, was inferred to be at least 6 times heavier than the Sun, thereby becoming the first black hole candidate. About 20 binaries are now known to contain an X-ray emitting compact object too massive to be a neutron star. (As of this writing, the record for the most massive – and most accurately measured – black hole candidate is 15.7 solar masses, for an edge-on X-ray binary in the nearby spiral galaxy M33.) This does not prove that these objects are black holes but, as Edwin Salpeter put it when the identification was still controversial, "A black hole in Cygnus X-1 is the most conservative hypothesis."

Finding 20 convincing black hole candidates among X-ray binaries, with another 20 probable ones based on their spectra and variability, is hardly a small harvest. There are on the order of 100 million neutron stars in our Galaxy, yet maybe only one in a million formed (and managed to remain) in a close binary and is currently accreting mass from its companion. A conservative guess would be that at least 1 percent of all massive stars are heavy enough to collapse to black holes. Thus, it would not be surprising if only a few percent of X-ray binaries contain a black hole stellar remnant, with the vast majority of such remnants moving solo through interstellar space, unseen. By these standards, our record of discovery is excellent.

However, the observations have not yet offered any signature that allows us to test whether the geometry of spacetime around these objects (which determines the flow pattern of infalling gas, etc.) is precisely that of a Schwarzschild or Kerr black hole, as Einstein's theory would predict. In principle, the spacetime geometry could be probed if the observational data were of high enough quality, and if the relevant gasdynamics and radiation processes could be modeled with sufficient accuracy. Such testing is in its infancy – in Chapter 10 we will discuss the first tentative steps.

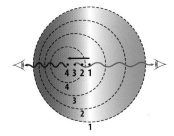

The Doppler effect. If a moving object (like the source depicted here, moving from position 1 to position 4) emits a succession of waves or pulses of light, the waves (represented here by circles numbered 1 to 4) will be more closely bunched together (and will arrive at higher frequency) toward the direction of motion; they will be more spread out (and have lower frequency) in the opposite direction. A source of light appears bluer (and brighter) if it is approaching and redder (and fainter) if it is receding. Measuring the Doppler shift of a spectral line of known frequency gives an accurate way to measure the approach or recession velocity of an astronomical object.

The most obvious signature of a black hole would be that of the horizon itself. All told, about half of the gravitational potential energy released by accretion will be radiated away throughout the disk. What happens to the other half depends on the nature of the compact object at the center. You may have been struck by how often the words "neutron star" and "black hole" have appeared together in the preceding pages. Unless there is an obvious giveaway such as pulsations, it is hard to distinguish an accreting black hole from an accreting neutron star just by observing its X-ray emission. But there is one important difference in the way they absorb matter that is crucial for detecting stellar-mass black holes in X-ray binaries. In contrast to a white dwarf or neutron star, a black hole has no "hard" surface. By the very definition of the black hole horizon, no energy is radiated as gas crosses into the hole. In fact, angular momentum can no longer keep gas swirling around a black hole when the gas is within three times the radius of the horizon (for a nonrotating hole; the corresponding distance is somewhat smaller if the hole is rotating). Any gas that spirals down to the radius of the "innermost stable orbit" can drop into the hole without further need of friction. Thus, an accreting black hole swallows whatever fraction of the gravitational potential energy is not radiated by the disk. If gravitational energy were not radiated from all radii in an accretion disk, but just emerged from the center, then accreting black holes would be undetectable. In a white dwarf or neutron star, the energy left over at the center makes a "splash" as the gas hits the stellar surface. We would be able to observe this "boundary layer" even if the accretion disk itself were invisible. But in the case of a black hole, the disk is all that we can see. It has long been a hope of astronomers that we would learn how to distinguish the radiation from the boundary layer from that of the disk on purely observational grounds. This is an active area of research, and a few leads look promising. But so far the only surefire signature of a black hole in an X-ray binary is its mass.

The black holes we have discussed so far – and the ones for whose existence the astrophysical arguments are most straightforward – are those with masses of a few suns and radii of approximately 10 kilometers. These represent the final evolutionary state of massive stars. But the black holes with the most robust (and diverse) observational signatures are millions of times more massive. In the next chapter we discuss the settings of such holes – the nuclei of galaxies. We shall see that conditions in a galactic nucleus may be propitious for the accumulation of a very large amount of mass, which could perhaps collapse to form a black hole. We will introduce the nearest of these "supermassive" black holes – the one at the center of our galaxy, the Milky Way – and set the stage for our subsequent investigation of the black holes that apparently inhabit the nuclei of all large galaxies.

4 GALAXIES AND THEIR NUCLEI

Thomas Wright's Milky Way. From inside the slab of stars an observer would see starlight as a band on the sky.

Over four hundred years have passed since Copernicus dethroned the Earth from the privileged central position accorded to it by Ptolemy's cosmology, and described the general layout of the Solar System as it is accepted today. But the eventual abandonment of Copernicus's helio-centric picture – in which the Sun is regarded as the central body in the Universe, not just the focus of the Solar System – and the full realization of the scale of the cosmos came about much more gradually and was completed only recently. In 1750, Thomas Wright of Durham, a self-educated teacher of astronomy, argued that the Milky Way, which appears to the naked eye as a diffuse band of light on the sky, could be explained as a thin slab or disk of stars in which the Solar System is embedded; William Herschel put this idea on a more precise footing in the 1780s. Wright's prescient ideas were popularized (in somewhat distorted form) through a pamphlet written by the philosopher Immanuel Kant, who added his own, equally remarkable conjecture that the Milky Way, our Galaxy, is just one fairly typical galaxy similar to the hundreds of millions of others that can be registered by a large telescope, and that galaxies are the basic units making up the large-scale Universe. Despite this startlingly modern view, the "spiral nebulae" photographed by late-nineteenth and early-twentieth century telescopes were not widely regarded as distant "island universes" until Edwin Hubble succeeded in establishing their enormous distances in the 1920s.

Our foray into the study of galaxies aims squarely at their centers – their *nuclei* – where we will find the conditions ripe for the formation of very massive black holes. Before focusing on the pyrotechnics sometimes engendered by these holes, and the strong evidence for their existence in "normal" galaxies like out own, we need to understand their environments and why they exist at all. To reach that point, we must first describe how galaxies are put together and how they evolve with time.

Normal Galaxies

Galaxies – giant assemblages of stars, gas, and dust – are supported against gravitational collapse through the motions of their stars. Beyond this general statement of gravitational equilibrium, no theory analogous to the well-developed theory of stellar structure exists to explain the gross properties of galaxies. Nevertheless, galaxies constitute the most conspicuous large-scale features of the cosmos. Only through diligent observation have we learned that galaxies possess fairly standardized properties, from which we can infer their structures and something about their fates.

The most familiar types of galaxy are the disklike, or spiral, systems. On a photograph, such a galaxy seems to be composed essentially

Edwin Hubble, shown here at the prime focus of the 200-inch telescope on Mt. Palomar, demonstrated that the galaxies seen through telescopes are distant "island universes," and discovered that the light from each one arrives with a redshift that is proportional to its distance from us – the first evidence that we live in an expanding Universe.

of stars and gas concentrated in a circular disk. A typical spiral galaxy contains about 10^{11} stars, and about 10 percent as much material in the form of gas, spread through a region about 50 000 light-years in radius. The stars and gas execute approximately circular orbits about the center of the disk, with centrifugal force balancing gravity. A typical star would take about 200 million years to orbit the center of the galaxy, at a speed of 200 kilometers per second; such a star would have completed 50 orbits since the time the typical galaxy formed. If our own Milky Way were viewed by a distant observer, it would appear disklike, the Sun lying about two-thirds of the way out toward the edge of the visible disk. The Andromeda galaxy, our nearest major neighbor in space, is about 2 million light-years away. Like other spiral galaxies, it contains stars of all ages, as well as gas. The Andromeda galaxy appears to be somewhat larger than the Milky Way; spiral galaxies in general have a spread of at least a factor 10 in mass.

In any galaxy classification scheme the main distinct class apart from the disk galaxies must be the "ellipticals." These are amorphous swarms containing between 10^8 and 10^{12} stars, each star tracing out a complicated path under the influence of the overall gravitational field. The smoothed-out appearance of an elliptical galaxy indicates that the orbits have become thoroughly randomized. An equilibrium has been established such that the tendency of gravitation to pull the system together is balanced by random motions, which by themselves would cause the galaxy to fly apart. Even though a typical photograph does not resolve the individual stars, over virtually the whole galaxy these stars are so widely spaced relative to their actual physical sizes that there is no significant chance of collisions, or even of close encounters, during the entire 10-billion-year lifetime of such a system (the same is also true of disk systems). Only in the densest parts of galactic nuclei is there some risk of stellar collisions; we shall see later that such encounters may contribute to the formation of black holes. Elliptical galaxies, generally less photogenic than spirals, come in a variety of apparent shapes ranging from nearly circular to highly flattened. The variety of shapes that we observe can be attributed partly to the different orientations of these galaxies relative to our line of sight, but also partly to genuine differences in shape.

Astronomers have developed elaborate taxonomies for classifying galaxies according to their shapes. The first such scheme was devised by Hubble, who (mistakenly) imagined an evolutionary sequence going from ellipticals to spirals. More recently, Gérard de Vaucouleurs formulated a scheme with over a hundred categories, and even this classification failed to include all "peculiar" systems! Most galaxies can be crudely categorized as either "disk" or "elliptical," but even a disk galaxy often contains an elliptical-like "bulge" in its center, while some ellipticals show vestigial disks. A slightly more sophisticated classification is based on the relative prominence of disk and elliptical components. The Sombrero Galaxy displays an average combination of disk and bulge components; most other disk galaxies are basically similar except that the bulge is either more or less prominent relative to the disk.

Disk galaxies contain cool gas and young stars, but elliptical galaxies contain neither. These seemingly incidental facts yield powerful clues to the nature of these systems. Stars are fueled by nuclear reactions in their interiors, and spend most of their time on the so-called main sequence, drawing their energy from the conversion of hydrogen into helium. A star like the Sun can remain in this phase for about 10 billion years. On the other hand, more massive stars (which can be identified observationally by their bluer colors, indicating a hotter surface layer) burn up their fuel at a much greater rate. They consequently have shorter lifetimes and eventually explode in violent supernova explosions,

Most large galaxies can be classified according to appearance, as either ellipticals or spirals. Spirals are divided into "barred" and "unbarred" varieties, and "irregular" galaxies form a separate class (not shown). Relative sizes of galaxy images reflect their distances more than their intrinsic dimensions.

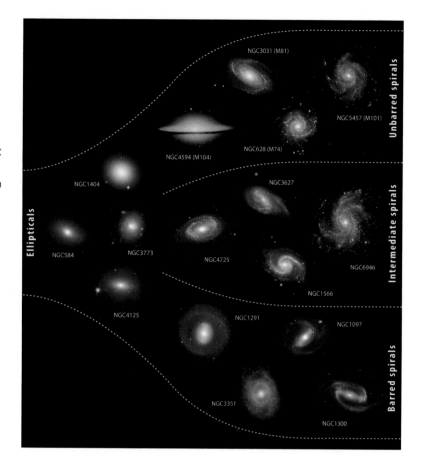

The "Sombrero Galaxy," some 30 million light-years away, is classified as a spiral because of its disk of stars and gas. It sports a prominent central bulge of stars, which, if seen in isolation, would resemble an elliptical galaxy. This image combines a picture taken by the *Hubble Space Telescope* in visible light (blue) with a near infrared exposure from the *Spitzer Space Telescope* (pink). Here the disk is outlined mainly by the infrared emission of warm dust; it also shows up as starlight in the *Hubble* image.

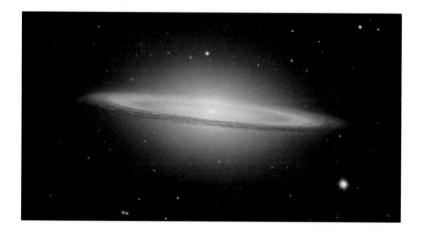

The dark linear feature, centered in its comet-shaped trail, is a "proto-planetary disk," or "proplyd" for short. This structure, located in the Orion Nebula, is believed to be an edge-on disk of gas in which a star and its retinue of planets are forming before our eyes.

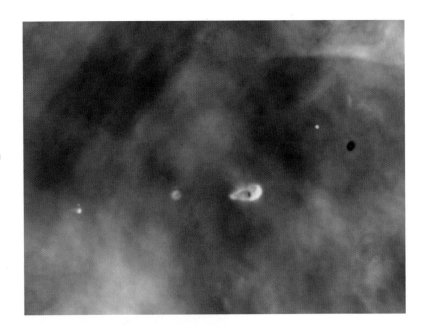

as we discussed in Chapter 2. In places such as the Orion gas cloud, we can see the actual process whereby gas is being converted into stars; and we can also see large numbers of blue stars that must have formed very recently in the history of the galaxy. When these stars die, they will eject much of their material back into the interstellar gas, from which new generations of stars can subsequently condense. The cool gas and young stars are different stages in a recycling process, and it should therefore not surprise us that the kind of galaxy that lacks cool gas also lacks young stars.

In their classic book on galactic dynamics, James Binney and Scott Tremaine assert that galaxies are to astronomy what ecosystems are to biology. The process of star formation, evolution, death, and rebirth going on in our Galaxy is, in a sense, an ecological cycle in which diffuse gas and stars share a symbiotic relationship. The ecological analogy reflects other features of galaxies: their complexity, ongoing evolution, and relative isolation. The cycling of matter between stars and gas is not perfect, however. Part of the gas that turns into stars becomes permanently trapped, in the sense that it becomes incorporated into long-lived (low-mass) stars, white dwarfs, neutron stars, or black holes. In galaxies with vigorous star formation, matter can be lost altogether in powerful winds driven by repeated supernova explosions. And galaxies can also acquire gas and stars by merging with smaller satellite galaxies that venture too close. The Milky Way should be thus refreshed in a few billion years when it swallows the Magellanic Clouds.

Gas incorporated into stars is recycled when the stars explode as supernovae or emit winds during the red giant phase of their evolution. Gas released during the red giant phase can be enriched in elements such as carbon, oxygen, and nitrogen; that released during a supernova explosion contains, in addition, a variety of heavier chemical elements. The matter remaining behind in white dwarfs, neutron stars, and black holes is taken out of circulation.

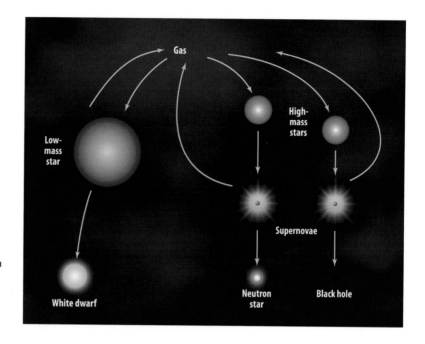

The Large Magellanic Cloud is an irregular galaxy in orbit around the Milky Way Galaxy, at a distance of about 170 000 light-years. Sites of active star formation glow pink with the light of ionized hydrogen.

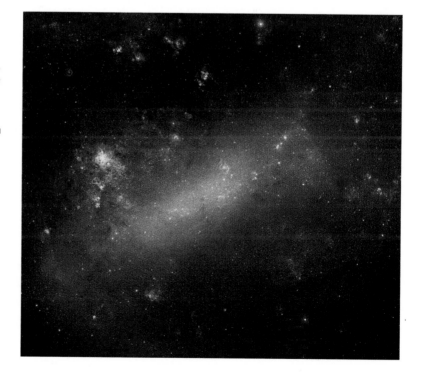

What determines whether a galaxy is an elliptical, with no young stars, or a star-forming spiral? Eddington claimed that a physicist on a cloud-bound planet could have predicted the properties of the gravitationally bound fusion reactors that we call stars. A star's radius, luminosity and temperature are more or less determined by its mass and internal chemical composition. We have not yet found an analogous set of arguments that can be used to "predict" the properties of galaxies with anything approaching the same level of confidence. It seems that a galaxy's properties are mainly determined by how it formed and how its surroundings evolved. To achieve a full explanation of galaxies requires setting them in a cosmological context – the "seeds" that developed into galaxies must have been implanted in the early Universe.

We know from observations that galaxy formation was well underway when the Universe was a small fraction of its current age of approximately 13.7 billion years. In the expanding Universe, looking to high "redshifts" not only corresponds to peering across vast distances – it also equates to looking back in time (see the box on pages 80–81). The most distant galaxies detected to date lie at redshifts greater than 6 (and possibly as great as 10), corresponding to an epoch less than a billion years after the big bang.

In the past decade there has been great progress toward developing a theory of galaxy formation and evolution. Though many details are very poorly understood – particularly in the latter stages, once stars begin to form – we are able to follow tiny primordial density fluctuations as they condense and coagulate to form galaxy-size masses. But our modern understanding of galaxies depends on an embarrassing fact: that most of their mass, probably more than 80 percent, is unaccounted for – it isn't in the stars and gas that we see, but takes some unknown "dark" form. This dark matter drives the formation of galaxies, and to a large extent determines their gravitational fields – therefore it affects the accumulation of matter all the way in to the central regions.

Dark Matter

Perhaps nothing confounds our understanding of galaxies so much as the realization that the gas and stars we observe constitute only a small fraction of the total mass of all large structures in the Universe, from galaxies on up. The observed stars and gas may be little more than tracers for the material that dominates the gravitational fields of these structures. The evidence for "dark matter" dates back more than seventy years, to the time when Fritz Zwicky realized that the aggregate mass of all the galaxies we could see in the Coma Cluster was inadequate, by a wide margin, to prevent the system from flying apart. The case for the existence of dark matter is now quite compelling; what dark matter consists of, however, remains a mystery.

Our own Galaxy and other spiral galaxies like the Andromeda Galaxy apparently have extensive haloes of dark matter, extending well beyond the outermost stars. One line of evidence for these dark matter haloes comes from mapping the speeds at which the disk rotates at various distances from the galactic center. Clouds of neutral hydrogen, themselves just a small fraction of the total mass, serve as tracers of the orbital motion. In spiral galaxies that are observed almost edge-on (the Andromeda Galaxy is the closest example), radio astronomers can infer the speed of the gas from the Doppler shift of the 21-centimeter spectral line, explained in the box on page 85. Moreover, this gas can be observed at radii far outside the edge of the conspicuous stellar disk. If the outermost hydrogen clouds were feeling just the gravitational pull of what we can see, their speeds should fall off roughly as the square root of distance outside the visible limits of the galaxy: the outer gas would move more slowly, just as Neptune and Pluto orbit the Sun more slowly than the Earth does. Instead, the velocity remains almost constant, implying that the outer gas is "feeling" the gravitational pull from a larger mass than the inner gas. It is as though we found that Pluto was moving as fast as the Earth, even though it is much farther from the Sun; if that were the case, we would have to postulate an invisible shell of matter, outside the Earth's orbit but inside Pluto's. Measurements like those shown on page 86 for Andromeda have been made for many disk galaxies. In some galaxies the inferred mass keeps increasing out to distances larger than 250 000 light-years, implying that "dark" matter contributes at least five times the mass of the "luminous" disk.

Further evidence for dark matter within our own Galaxy comes from measured velocities and inferred orbits of dwarf satellite galaxies and outlying globular clusters found in our Galaxy's halo. The total mass of the Local Group of galaxies (of which our Galaxy and Andromeda are the dominant members) can be derived from the so-called timing argument, whereby one calculates how much mass would be needed in order to bring about the present situation in which Andromeda and our Galaxy, now 2 million light-years apart, are observed to be "falling" toward each other. These estimates suggest that the two haloes may be so massive that they extend out until they effectively merge.

Similar conclusions hold for elliptical galaxies, though here the evidence is different. These galaxies contain gas so hot that its pressure prevents it from condensing into the center (or into a disk). This gas is detected by its X-ray emission. Moreover, the X-rays tell us how the temperature of the gas varies with radius. Even far out, the gas is so hot that

The Expanding Universe

The expansion of the Universe has been known since the 1930s, when Edwin Hubble realized that distant galaxies are receding from us at speeds proportional to their distances. His discovery proved so momentous that it has received the designation Hubble's law.

Although we see galaxies moving away from us no matter which direction we look, this doesn't imply that we are in a special central position. To see why, consider the lattice in M. C. Escher's *Cubic Space Division*. Imagine that the rods in this lattice all lengthened at the same rate: then the vertices would all move away from one another, and an

M. C. Escher's infinite lattice has no center. If the rods all lengthened at the same rate, an observer at any vertex would see the others receding according to Hubble's law.

observer on any one of them would see the others receding. Moreover, every time a neighboring vertex receded by 1 unit of length, successively more distant vertices would recede by 2, 3, 4, . . . units of length. In other words, the vertices recede in accordance with Hubble's law. This is a good "broad brush" analogy for the expanding Universe, except that the galaxies are not laid out in a regular lattice. Actual galaxies are grouped into clusters and superclusters, and consequently move locally under the gravitational influence of other cluster members. The galaxies within a cluster are not receding from one another: Hubble's law shows up clearly only on scales larger than clusters.

The light from very remote objects set out when the Universe was at an earlier stage in its expansion and galaxies were more closely packed together. Because of light's finite speed, what we actually observe is better represented by another Escher picture, *Angels and Devils*: the scale of the pattern (i.e., the separations of the galaxies) is compressed as we look farther away. The wavelengths of light reaching us from an earlier era have stretched by a factor equal to that by which the Universe has expanded since the light was emitted. Since light with a longer wavelength is "redder," the radiation we receive from distant galaxies

is said to be "redshifted." The amount of the redshift can be associated with the speed of recession, and we can then use the calculated recession speed to derive the distance to a remote galaxy, using Hubble's law.

Astronomers normally quote redshifts in terms of the number z, defined such that the received wavelength is $(1 + z)$ times the emitted wavelength. For example, the redshift of the quasar 3C 273 (Chapter 5) is 0.16, meaning that the Lyman alpha line of hydrogen, which was emitted with a wavelength of 1216 Å, reaches our telescope with a wavelength of 1410 Å, or a factor of 1.16 times longer. For the (relatively nearby) galaxies that Hubble himself observed, z was very much less than 1. However, galaxies – and especially the luminous quasars and radio galaxies that we will meet later – are now routinely observed with redshifts exceeding 6. The number $(1 + z)$ tells us how much the Universe has expanded while the light has been on its journey. The light from an object with $z = 3$, for instance, set out when the Universe was compressed by a factor $(1 + 3) = 4$, and was denser by a factor $4^3 = 64$.

If the Universe expanded at a uniform rate, then when it was 4 times more compressed it would have been 4 times younger. However, the actual history of expansion depends on the Universe's contents. The mutual gravitational pull of all normal forms of matter and energy tends to decelerate the expansion; this effect seems to have dominated cosmic evolution at redshifts higher than 1 or 2. Since the early stages of expansion were faster, the Universe at $z = 3$ was less than 1/4 its present age. But several different kinds of observations from the past decade have revealed an astonishing feature of the recent expansion: it seems to be *accelerating*. The physical cause of the acceleration is unknown: it is usually attributed to a hypothetical *dark energy* that produces a repulsive form of gravity. This dark energy, if it exists, constitutes about 73 percent of the Universe, with all forms of matter (mostly dark matter) and radiation making up the remaining 27 percent.

To determine actual distances, we need to know the so-called Hubble constant. When an object's redshift is small (and z much less than one) its distance is just z divided by the Hubble constant. To determine the Hubble constant, one needs to make accurate measurements of the distances to a sample of galaxies, using measurements *other than their redshifts*. Obtaining these measurements is a difficult procedure, but we

are confident that we now know the Hubble constant to within plus or minus 5 percent. Thus, we can state that 3C 273, with its redshift of 0.16, lies about 2 billion light-years from us. The light now reaching us from 3C 273 was emitted about 2 billion years ago (actually slightly less, because of the stretching of space that has occurred in the intervening time).

The relationships among age, distance, and redshift are more complicated for high-redshift objects. The light now reaching us from high-z objects set out when the Universe was substantially more compressed and presumably not expanding at the same rate as it is now. Most cosmologists have adopted the so-called "concordance," or ΛCDM, model for cosmological evolution. The Greek letter Λ (lambda) represents the dark energy component of the Universe, here treated as Einstein's "cosmological constant," while the letters CDM stand for "cold dark matter," a description of the principal matter component. According to the ΛCDM model, the Universe is 13.7 billion years old and galaxies or quasars with redshifts greater than 6 are seen as they were less than a billion years after the big bang. When we quote ages of high-z objects in this book, we use ΛCDM values. For quick estimates, cosmologists often talk loosely of the

"Hubble radius." This is the distance we would calculate for an object with $z = 1$ if the simple proportionality between redshift and distance applied out to such a large redshift. This distance, divided by the speed of light, gives an equally rough "age" for the Universe – which turns out to be surprisingly close to the ΛCDM value.

Because of light's finite speed, we do not actually see a uniform lattice: distant parts of the Universe are seen as they were in the remote past. When telescopes probe regions at high redshifts (near the observational "horizon" – equivalent to the regions near the boundary of Escher's *Angels and Devils*), they reveal a phase of cosmic history when everything was packed more closely together.

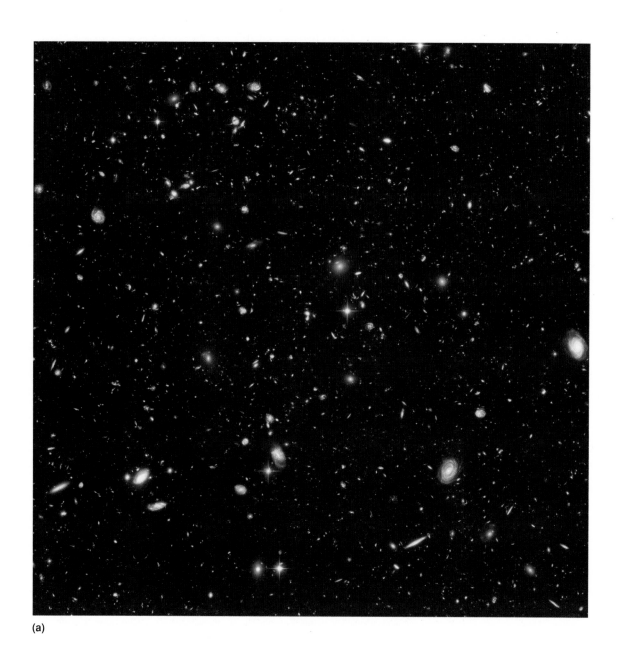

(a)

The Hubble Ultra Deep Field – a patch of sky only one-tenth the width of the full Moon – contains images of at least 500 high-redshift galaxies in existence less than a billion years after the big bang. The galaxies have very blue colors (after correcting for the redshift), implying that they are forming stars at a very high rate compared to modern galaxies. These distant galaxies are so faint that they are not readily visible on the image (a) at left – magnified images of 28 of them are shown at right (b).

(b)

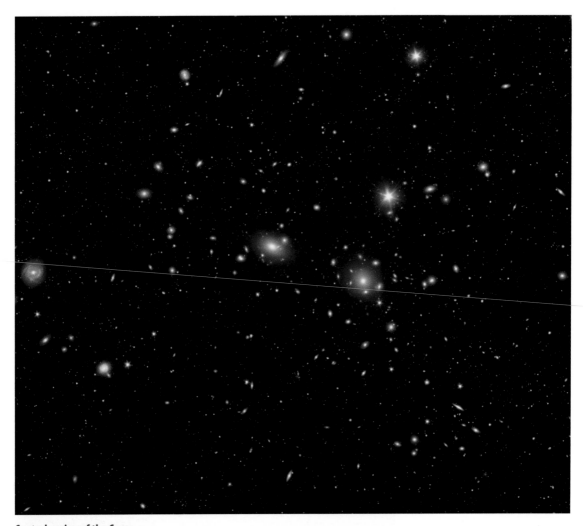

Central region of the Coma Cluster. In this system there must be several times more mass in some "dark" form than in the stars and gas we detect – otherwise the galaxies would fly apart and the cluster would disperse. This mosaic combines visible-light data from the Sloan Digital Sky Survey with infrared images from the *Spitzer Space Telescope*.

it wouldn't be bound to the galaxy at all unless it were being held down by a stronger gravitational force than the stars alone provide. So elliptical galaxies must have extended dark haloes, just as disk galaxies do.

Clusters of galaxies appear to have even larger concentrations of dark matter, compared to the luminous matter forming the galaxies. Three different kinds of observations reveal the presence of dark matter in clusters. The first is a refined version of Zwicky's original method. For clusters to be in equilibrium, there must be a balance between the effect of the relative motions of the galaxies (which tends to disperse the cluster) and the effect of gravity (which, if the galaxies weren't moving, would cause them to fall together at the cluster center). The speeds of the galaxies (or at least the component of their motion that is directed

The 21-Centimeter Line of Hydrogen

Explaining the hydrogen spectrum – the Lyman lines, the Balmer lines, etc., discussed in the box on pages 38–39 – was one of the triumphs of the early quantum theory in the 1920s. But some complications were appreciated later. One subtle effect, which turns out to be especially important for astronomy, depends on the fact that each electron and proton has an intrinsic "spin." Just as a charged spinning ball, or a loop carrying an electric current, creates a magnetic field, a charged particle with spin behaves like a tiny magnet.

Two magnets placed side by side repel each other when they are aligned, and attract when they point in opposite directions. An electron "in orbit" around a proton is similarly pushed slightly farther out if the magnetic fields are aligned than if

they are misaligned. Energy can therefore be released if the electron "flips" so that it becomes magnetically misaligned.

The effect is tiny: only a small quantum of energy is released when an electron's magnetic field flips from being aligned to being misaligned with the proton's. These so-called hyperfine transitions are therefore signaled by radiation of very low frequency – in other words, of very long wavelength. This radiation can be detected by radio astronomers who tune their receivers to a wavelength of 21.1 centimeters.

Radio telescopes are sensitive enough to pick up the "21-centimeter line" from hydrogen in interstellar clouds – indeed, this proves to be the best technique for

probing diffuse hydrogen in other galaxies as well as our own. Another feature of the line is that it is very sharp: a radio receiver has to be tuned exactly right to pick it up. The wavelength will of course be stretched by the Doppler effect if the gas is moving away from us; the extent of the Doppler shift tells us the speed of the gas. Since radio telescopes can routinely measure the wavelength to one part in 100 000, speeds can be measured with a precision of 1/100 000 the speed of light – to within 3 kilometers per second. The 21-centimeter line can therefore be used to map out how gas is moving within galaxies. Astronomers can infer the strength of the gravitational field that causes these motions; this technique provides one of the main lines of evidence for dark matter.

along our line of sight) can be measured from the Doppler effect. Once we know the motions, we can calculate what the cluster's total mass must be. This is a simple application of the "virial theorem," discussed in the first chapter, which asserts that the gravitational binding energy – which is related to the mass – must equal the internal kinetic energy if a system is in equilibrium. This method of calculating a cluster's mass is complemented by X-ray studies – analogous to those performed for elliptical galaxies – of the hot gas that fills the space between galaxies in clusters.

The third technique yields the most dramatic evidence for dark matter, and also enables astronomers to trace how the mass is distributed throughout a cluster. The method makes use of the deflection of light rays by the cluster's gravitational field. That the path of a light ray is deflected by a gravitational field is one of the principal predictions of general relativity, and the first one to be confirmed observationally. This effect distorts the images of objects lying behind the cluster in the background, very much as a glass distorts an image – hence the phenomenon is often referred to as "gravitational lensing." Some remarkable arclike objects are observed, which are highly stretched and magnified images of faint galaxies lying

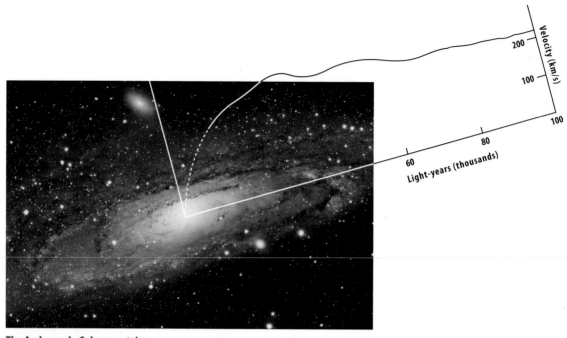

The Andromeda Galaxy contains a spinning disk viewed almost edge-on. By observing the 21-centimeter line of hydrogen, radio astronomers can detect cool gas in the disk out to distances far beyond the apparent optical boundary of the galaxy. The orbital speed of the gas, which can be inferred from the Doppler effect, is plotted on the graph. The fact that the orbital speed remains high even at the largest distances indicates that the outermost gas must be "feeling" the gravitational pull of extra matter outside the visible part of the galaxy. Similar evidence exists for many other galaxies, and suggests that very extended "dark haloes" contain several times more mass than all the visible stars and gas in galaxies.

far behind the cluster. More detailed studies reveal that many background galaxies viewed through rich clusters are somewhat distorted, though less dramatically than the arcs. Such data in principle allow astronomers to reconstruct the two-dimensional projection of the cluster's mass density on the sky. This line of evidence would have delighted Zwicky: not only a pioneer investigator of dark matter, he was also one of the first to speculate about gravitational lensing by galaxies.

All three techniques are more or less in agreement, and imply a discrepancy of about a factor of 5 to 10 between the total mass of galaxy clusters and the sum of the masses of the stars and gas in the individual galaxies, plus the gas between the galaxies. That is, there seems to be at least 5 times more dark matter in the Universe than luminous matter. The measurements also agree with other, less direct methods for estimating the total matter content of the Universe – by studying the distribution of galaxies on large scales, analyzing slight distortions in their shapes across broad swaths of sky ("weak lensing"), and mapping tiny fluctuations in the temperature of the *cosmic background radiation*, the weak glow left over from the big bang. Measuring the total amount of matter is of fundamental importance for cosmologists, since it determines the tendency of the cosmic expansion to decelerate. But even with a ratio of 5-to-1 between dark and luminous matter, the total matter content is less than 30 percent of

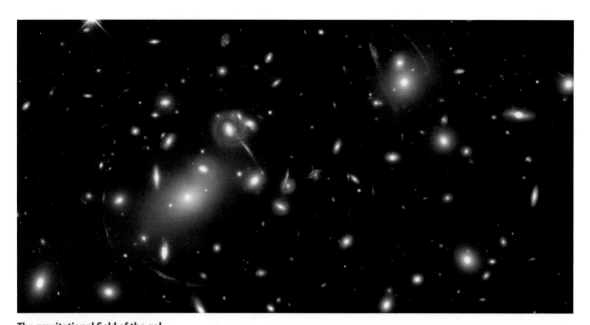

The gravitational field of the galaxy cluster Abell 2218 distorts the images of more distant galaxies, stretching them into concentric arcs. By measuring the pattern of distortion, it is possible to map out the distribution of matter – dark as well as luminous – in Abell 2218. The results show that this cluster is heavily dominated by dark matter. The longest arcs are magnified more than ten times. The cluster is therefore acting like a "telescope," allowing astronomers to study galaxies at high redshifts (up to at least 2.5, and possibly much higher) that would otherwise be invisibly faint.

the "critical density" above which the expansion of the Universe would eventually grind to a halt, and reverse. The remainder appears to be in the form of a mysterious "dark energy" that is causing the expansion to accelerate (see box on pages 80–81).

What is the dark matter made of? All we can state with confidence is that it must be mainly in some unknown form – it can consist neither of ordinary stars nor of the gas that emits X-rays. Whatever its composition, it must be almost completely non-interactive in all ways except gravitational. This key feature of dark matter is illustrated vividly by the "Bullet cluster," thought to be the aftermath of the collision between two clusters of galaxies. The X-ray emitting gas particles in each cluster interact electromagnetically, as well as gravitationally, and pile up in a heap near the midpoint of the collision; whereas the dark matter haloes of the clusters, along with the galaxies, pass through each other.

One shouldn't be too surprised that most of the matter in the Universe is dark: there is no reason why everything should shine, and it isn't difficult to think of possible candidates for the makeup of the dark matter. Indeed, the problem is rather to decide among a long list of possibilities. Obviously, direct detection of the dark matter itself would be the "cleanest" and most decisive discriminant: low-mass stars and planetlike objects in the Galactic halo, too faint to be seen directly, would reveal themselves by deflecting and magnifying the light from more distant stars; exotic particles that pervade the Galaxy (and therefore are continually passing through every laboratory) might be detected by sensitive experiments.

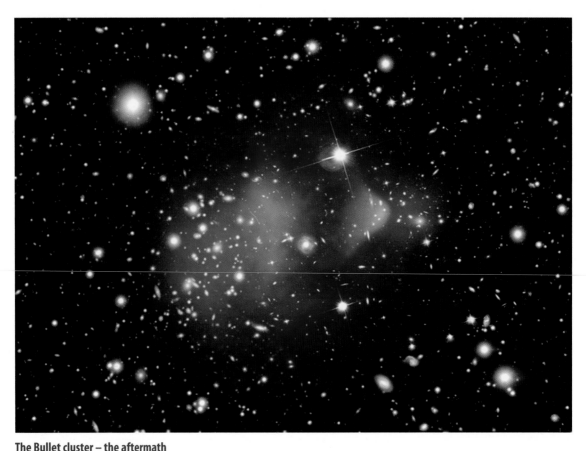

The Bullet cluster – the aftermath of a collision between two clusters of galaxies – proves the weakly interacting nature of dark matter. The distribution of dark matter (blue), mapped using gravitational lensing, is superimposed on an image of the colliding clusters, while the X-ray emitting gas is shown in pink. Whereas the gas particles collided and became trapped between the clusters, the dark matter haloes passed through each other.

The most "conservative" guess is that the dark matter in our Galaxy is in the form of stars too low in mass to ignite nuclear fuel in their centers. The masses of these so-called brown dwarfs or "Jupiters" would have to be below 7 percent of a solar mass in order for these objects to be faint enough to have escaped detection by conventional astronomical techniques. Even if they emit no radiation (apart, presumably, from very faint infrared emission), the objects would still disclose themselves by occasionally magnifying the light from a background star, a by-product of the gravitational lensing effect. The strength of the magnification would depend on how far the dark foreground star is from Earth, compared to the background source, and the precision with which the dark object is "lined up" in front of the source. Because the mass of a brown dwarf is so small, the alignment has to be near perfect in order to obtain an observable effect. Brown dwarfs in the halo of our Galaxy would create magnified images of the more distant stars in the Magellanic Clouds with an apparent angular size of a few microarcseconds (10^{-6} arcsecond), far too small to be resolved with existing telescopes.

Gravitational "microlensing" occurs when a compact object lies along the line of sight to a star. If the object is exactly in front of the star, the latter's brightness is enormously magnified; observing with very sharp resolution, we see the star distorted into a ring whose radius is the "Einstein radius," r_E. If the alignment isn't quite perfect, two highly magnified images, separated by $2r_E$, are seen. Significant (though less extreme) magnification occurs whenever the offset is less than r_E.

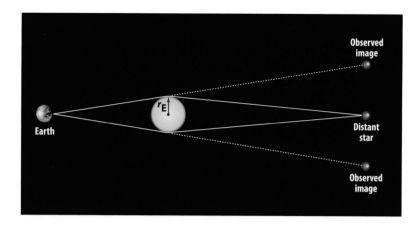

(A telescope with a resolution of a microarcsecond would enable us to read a newspaper on the Moon.) One can nevertheless identify instances of lensing by watching for the characteristically symmetric rise and fall in magnification that is visible when a halo object moves almost across the line of sight to a background star.

Even if low-mass stars contributed all of our halo's dark matter, the probability of there being significant lensing, at any instant, along a typical line of sight would be only 3 chances in 10 million. (This result does not depend on the typical mass of the "brown dwarfs.") To stand

This graph shows how the total magnification of a star rises and falls when a compact object moves steadily across the line of sight to a background star. Curves are shown for cases when the foreground object passes exactly in front of the background star, and also when the closest alignment is 0.2, 0.4, 0.6, 0.8, and 1.0 times r_E (top right). The timescales correspond to an event in which a star in the Large Magellanic Cloud is lensed by a compact object of 0.01 solar masses in our Galactic halo, which moves with the typical velocity expected of halo objects. For different "lens" masses, the times would scale as the square root of mass.

Data for a candidate "microlensing" event in the Magellanic Clouds, observed by Charles Alcock and collaborators. The maximum magnification is about a factor of 7. The rise and fall is symmetrical, and follows the behavior expected. Further evidence that the brightening of the star resulted from lensing is that the red and blue curves track each other closely; gravitational lensing should be achromatic.

a chance of detecting this effect, one must monitor the lines of sight to millions of different stars. Several groups have succeeded in doing this within the past 15 years, and there are now hundreds of candidate lensing events. The results suggest that the contribution of "brown dwarfs" to our Galaxy's halo is *not* sufficient to account for most of the dark matter.

The remnants of a hypothetical population of massive stars, which might have formed early in galactic history, might also contribute to the dark matter in our Galaxy's halo. These remnants could be neutron stars or white dwarfs that have cooled off to the point of invisibility, or black holes with masses ranging up to 10^3 solar masses or higher. Such objects would be hard to detect by their lensing signatures because the individual events would be much slower and less frequent than events involving brown dwarfs. Several possible detections have been claimed, but these identifications are still controversial.

One feature that both brown dwarfs and stellar remnants share is that they are forms of "baryonic" dark matter – that is, matter composed of ordinary subnuclear particles such as protons and neutrons (or, in the case of black holes, resulting from the collapse of such matter). In the standard theory of the expanding Universe (the "hot big bang"), most of the helium, all the deuterium, and some of the lithium observed today were created by thermonuclear reactions during the first few minutes following the initial explosion. So far, no other theory has been proposed that comes close in its ability to produce all of these elements in roughly the correct quantities. In order for the reactions to yield the observed abundances, however, the baryon density must be between 3 and 10 percent of the total density of matter needed to "close" the Universe (that is, to cause it eventually to recollapse). Studies of the cosmic background radiation have refined this value to between 4 and 5 percent. Given this cap on the baryon density, it is doubtful that baryonic material could account for all dark matter inferred to exist in galaxies and clusters; it most certainly could not account for the amount of dark matter that seems to pervade larger structures known as "superclusters."

This argument has convinced most cosmologists that 80 to 90 percent of the dark matter in the Universe is of a "nonbaryonic" form, probably consisting of subatomic particles not ordinarily found in matter. The most obvious candidate is neutrinos, which have an "edge" over other nonbaryonic candidates in that they are at least known to exist. Neutrinos, which apparently come in three types, are particles much less massive than electrons. According to the standard big bang theory, there should be, on average, 100 million neutrinos of each type per cubic meter of space. Because neutrinos outnumber baryons by such a large factor (by more than 10^8, on average), they would not need to have a very high mass in order to have an important cumulative effect on cosmic dynamics: they would close the Universe if their mass were only one ten-thousandth that of an electron.

Unfortunately, the masses of neutrinos are unknown. There is evidence that they are not completely massless, but their masses could be so low that their effect on cosmic expansion is negligible. In fact, if neutrinos dominated the dark matter density they would hamper the galaxy formation process. They would move with random speeds close to the speed of light, and would not respond to the gravitational attractions of galactic haloes or other structures. To make matters worse, they would tend to erase existing seeds of cosmic structure before the seeds had a chance to grow. Thus, even if neutrinos contribute to the dark matter density, there must be a dominant component of dark matter particles that move much more slowly.

Theoretical physicists have a long shopping list of hypothetical particles that might have survived in the requisite numbers from the early Universe. One widely discussed possibility goes by the whimsical name of WIMPs, which stands for "Weakly Interacting Massive Particles." (Not to be outdone, the proponents of dark stars and stellar remnants call their targets MACHOs, for "Massive Compact Halo Objects.") WIMPs could be comparable to protons in mass, or they could be much more massive. Several groups are attempting to detect WIMPs (and other candidate dark matter particles) directly, by measuring the recoil that would occur when one of these particles hits an atomic nucleus in a laboratory apparatus. However, even if WIMPs exist in sufficient numbers to provide all the dark matter in our Galaxy's halo, these collisions must be extremely rare. Moreover, the detection apparatus is bedeviled by extraneous collision events produced by the internal radioactivity of the equipment and by cosmic rays (energetic particles of normal matter from space, possibly produced in supernova explosions). Consequently, the experiments must be carried out deep underground.

Ingenious schemes for detecting a halo population of exotic particles seem among the most worthwhile and exciting high-risk experiments in physics or astrophysics. A null result would surprise nobody; on the other hand, such experiments could reveal the existence of fundamental new forms of matter as well as determine what more than 80 percent of the mass in the Universe consists of. The late Professor Roderick Redman of Cambridge, a "no-nonsense" observer with little taste for speculation, once claimed that any competent astrophysicist could reconcile any theory with any set of facts. An even more cynical colleague extended this claim, asserting that the astrophysicist often need not even be competent. Dark matter theorists have in the past all too often exemplified Redman's theorem, and its extension as well. But searches may soon yield definitive results; in any case, various constraints are now restricting the tenable options. Moreover, it is not "wishful thinking" to expect that there may be more than one important kind of dark matter – for instance, nonbaryonic dark matter could control the dynamics of large clusters and superclusters, even if individual galactic haloes contained a lot of brown dwarfs or stellar remnants.

It would be especially interesting if astronomical techniques were to reveal some fundamental particle that has been predicted by theorists, but not yet encountered in the laboratory. If such particles turned out to account for the dark matter, however, we would have to view the galaxies, the stars, and even ourselves in a downgraded perspective. Copernicus dethroned the Earth from any central position. Early in the twentieth century, Harlow Shapley of Harvard College Observatory and Edwin Hubble demoted the Sun from any privileged location in space. But now even "baryon chauvinism" might have to be abandoned: the protons, neutrons, and electrons of which we and the entire astronomical world are made could be a kind of afterthought in a cosmos where exotic elementary particles control the overall dynamics. Great galaxies could be just a puddle of sediment in a cloud of invisible matter 5 times more massive and extensive.

We should be pleased that we have been able to deduce the existence of dark matter, but sobered by our ignorance of its nature. The entities that collectively constitute most of the mass of the Universe may have individual masses anywhere in the range from 10^{-36} kilogram up to 10^{36} kilograms – an uncertainty of more than 70 powers of 10! Astrophysics may not always be an exact science, but seldom is the uncertainty as gross as this.

The range of options could be narrowed if there were less uncertainty about what particles might be created in the earliest stages of the expansion of the Universe. Computer simulations of galaxy formation have also yielded indirect (but very important) evidence on the nature of the dark matter – for example, limiting the permitted contribution of "hot dark matter," such as neutrinos. Since more than 80 percent of the gravitating material in a galaxy is "dark," the way in which galaxies form and cluster plainly depends on what the dark matter is. Under specific assumptions about conditions shortly after the big bang, one can compute how cosmic structure develops with time, and then compare the results with observations. The agreement is impressive, and has led to a promising, if incomplete, model for how galaxies may have formed and evolved.

Galaxy Formation

Galaxies are thought to have arisen from tiny density fluctuations in the Universe laid down by quantum mechanical effects, an instant after the big bang. After evolving passively for several hundred thousand years (to about a redshift of 1000), these fluctuations began to condense and coalesce, until they became well-defined structures at redshifts between 100 and 40, several tens of millions of years after the big bang. Since dark matter dominates over baryonic matter 5 to 1, it was the dark matter's gravity, primarily, that drew these condensations together against the overall expansion of the Universe. The baryonic gas went along for the ride, mimicking the evolution of these so-called "dark matter haloes."

This situation began to change at redshifts between about 40 and 20 (about 50 to 200 million years after the big bang), when gas acquired a means – the production of trace amounts of molecular hydrogen – to cool down and lose its internal pressure. The cool gas was drawn more deeply into the haloes, perhaps forming the first stars. Gas enriched with heavy elements would have been expelled from these star-forming haloes by the first supernova explosions. Such "metal enrichment" would have enhanced the ability of gas to cool; at the same time, the explosions themselves could have limited the formation of additional stars by expelling the available gas.

None of these complex gaseous processes would have had much effect on the evolution of the dark matter haloes, which can be modeled with confidence using computers. The galaxy formation process is thought to be *hierarchical*, with larger haloes forming through the coalescence of smaller ones. Large simulations like the *Millenium Run* make specific predictions about the clustering properties of haloes, their mass distribution, etc., which broadly agree with observational surveys.

Unfortunately, the gaseous processes – responsible for the luminous structures we can see – are much harder to model. In the earliest star-forming haloes, rotation may have been insufficient to keep gas from falling all the way to the center. Each of these systems might have formed a compact cluster of stars, or perhaps a single massive star, and the mergers of these haloes could have given rise to amorphous stellar systems resembling globular clusters or dwarf elliptical galaxies.

Larger haloes, especially those formed by "major mergers" – that is, mergers of similar-sized haloes – would tend to have substantial rotation. In these systems, most of the gas could not drift all the way into the center, but would collect in a disk – perhaps giving rise to the first spiral galaxies. Major mergers can also destroy disks: large elliptical galaxies probably formed through the collisions and mergers of similar-sized spiral galaxies.

Thus, we can understand qualitatively why those galaxies with no young stars should be ellipticals, whereas those where star formation is still going on should be disk-shaped. Only where gas can collect, and condense, is there a chance for star formation to proceed.

The Millenium Simulation Project, led by Völker Springel at the Max Planck Institute for Astrophysics in Garching, Germany, used 10 billion particles to simulate the development of cosmic structure in a cube 2 billion light-years on a side. The snapshots shown here represent the distribution of dark matter in the present-day Universe on three scales, indicated by the horizontal bars. Note the spongelike structure of the dark matter distribution on the largest scales, resolving into filaments as one zooms in. Large galaxies and galaxy clusters are believed to form preferentially at the intersections of filaments.

(a)

(b)

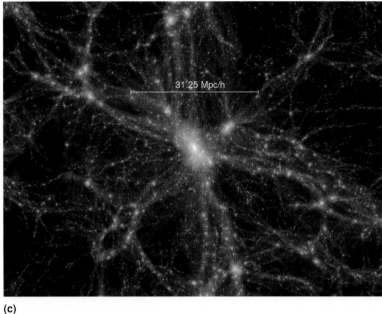

31.25 Mpc/h

(c)

We can also understand, roughly, why the Coma Cluster consists of distinct galaxies, instead of being a huge amorphous agglomeration of 10^{14} stars. Simple calculations suggest that a collapsing turbulent gas cloud would be unable to cool and fragment into stars if its mass exceeded 10^{12} solar masses, or if its radius were much larger than 300 000 light-years. Indeed, in large clusters of galaxies, such as Coma, most of the baryonic matter is in the form of hot, X-ray emitting gas.

As haloes merge, their average densities tend to decrease – they become more diffuse. But there is a countervailing tendency of galaxy formation that seems just as inexorable: at each stage, more gas settles toward the center of the merged halo, separating itself out from the dark matter. This is because the collapse of gas is highly dissipative, in the sense that colliding globules of gas produce shock waves that radiate away some of their energy of motion. Having lost energy that could have made them bounce apart from each other, the gas globules have a tendency to "stick together" and to accumulate in the deepest parts of the halo that rotation allows them to reach. Dark matter, on the other hand, is "weakly interacting," and has no way of radiating away its energy. Thus, it should be no surprise if the hierarchical process of galaxy formation, driven mainly by the gravitational pull of nonbaryonic dark matter, should lead to very dense galactic nuclei where baryonic matter rules.

In this snapshot from a simulation of primordial star formation by Tom Abel and collaborators, cold neutral hydrogen and helium gas is shown in blue and hot ionized plasma is shown in red. The region depicted is over 5000 light-years across. The red gas has been ionized by ultraviolet radiation from the single, massive star that has formed at the center of the image. As hot, pressurized material flows away from the star at speeds of up to 30 kilometers per second, it creates a shell-like structure in the red gas. Dense (blue) knots of material may be able to survive the short period of ultraviolet illumination and might seed further star formation.

Galactic Nuclei

If you examine a photograph of a galaxy, it is easy to see that the concentration of stars increases toward the center, the so-called nucleus. The nucleus of the Milky Way is obscured at visible wavelengths by a thick veil of dust and gas, but observations at infrared wavelengths have revealed as many as a million stars within the innermost single light-year. Compare this to the concentration of stars near us, which is about 0.006 suns per cubic light-year. On average, stars are 300 times closer together in the Galactic Center than they are in the vicinity of the Sun.

The nuclei of galaxies would appear to be rather dangerous places, in which stars are not only packed together but are also moving around randomly at speeds of up to several hundred kilometers per second due to their mutual gravitational attractions. The only other places where we know such high stellar densities to exist are in the centers of globular clusters, but there the random velocities are 10 to 100 times slower. (Some astronomers have speculated that galactic nuclei formed from the coagulation of globular clusters.) Violent collisions between stars are inevitable in galactic nuclei, and indeed are predicted to occur about once every 10 000 years in the nucleus of the Milky Way.

The fate of such a dense central concentration of stars is not easy to predict, and forms the core of a debate that started in the 1960s and still has not been resolved. Part of the problem is that the outcome

The merger of two spiral galaxies is depicted in this time sequence from a simulation by Lars Hernquist and collaborators. Each panel is approximately 370 000 light-years on a side and the simulation spans nearly 2 billion years. The brightness indicates the projected density of stars, while the color indicates that the gas content evolves from gas-rich (blue) to gas-poor (red). Note that the merger destroys the spiral structure, leaving behind a remnant that resembles an elliptical galaxy.

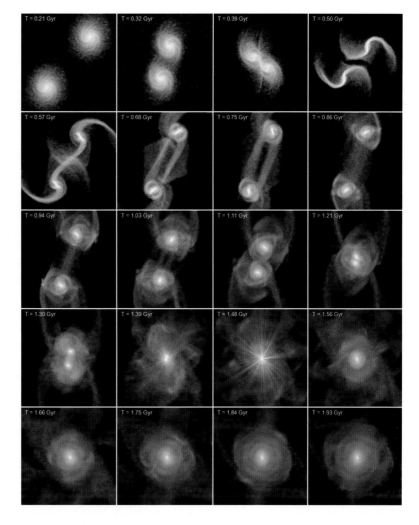

of even a single stellar encounter is not well understood theoretically. When considering the structure of a galaxy as a whole, it is acceptable to treat individual stars as "point masses" of infinitesimal size. But such an approximation will not do when considering encounters in which two stars pass within a few stellar radii of one another. Then the tidal forces deform the stars' internal structures, heating them slightly and placing a drag on their motions. If this drag slows their relative motions enough, the stars may fall into orbit about one another. In this way, two stars, initially moving independently of one another, can become locked in a "tidal capture binary." (The tidal capture mechanism probably accounts for most of the close binaries identified in globular clusters.) If the tides are strong enough, the stars may eventually spiral together and coalesce into a single, more massive star.

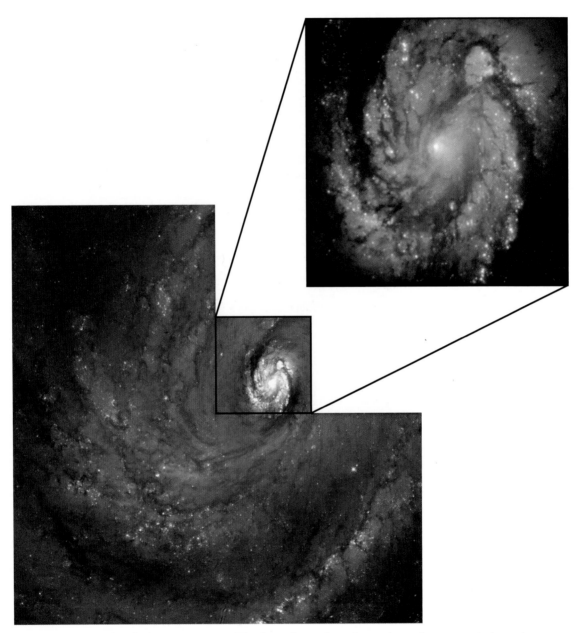

An intense concentration of star-light marks the nucleus of M100, a spiral galaxy in the Virgo Cluster.

The fate of a coalesced star depends on exactly how the merger takes place, and especially on the evolutionary state of the participants in the merger. Stellar evolution theory tells us how to predict the subsequent evolution of a star from some point in time, provided that we know the star's chemical *structure* at that point. This means that we need to know not only the total ratio of hydrogen to helium in the star, but also how that ratio changes with depth. If we start with a pure hydrogen star at the beginning of its life, we can follow the chemical changes step-by-step as nuclear reactions alter the composition. But what is the initial compositional structure of a star formed by coalescence? We can guess the overall composition by adding up the contents of each participant in the encounter, although mass may be lost during the coalescence process, making even this estimate uncertain. Variations of the composition with depth, however, will be affected by complicated and hard to predict mixing processes. Our best hope is to study the coalescence process "experimentally," using computer simulations. These are in their infancy now, but their sophistication is increasing rapidly.

Coalescence is the likely outcome for two stars encountering each other at velocities that are low compared to the escape speed from the surface of either star. The escape speed from the Sun, for example, is 618 kilometers per second. At much higher relative velocities, the stars literally smack into one another, releasing high-speed plumes of gas into space. Whereas low-speed collisions build up high-mass stars, high-speed encounters demolish the existing stars. In the latter case, as in the former, it is impossible to make detailed predictions of the outcome (especially the fraction of mass liberated) without computer simulations. Collisions are mainly disruptive when they take place in clusters of fast-moving stars having typical random velocities of more than 1000 kilometers per second. In the central star cluster of the Milky Way, the typical random speed is only about 200 kilometers per second, and so encounters between these stars should be in the coalescent regime. Indeed, our Galactic Center displays an unusually large abundance of massive blue stars, at least some of which could have been produced by coalescent encounters within the last million years or so.

Stellar encounters will certainly affect the evolution of a galactic nucleus, but we are still not sure exactly how. Some of the possibilities are illustrated in the flow chart on page 100. By itself, widespread stellar coalescence would tend to make the average star in the cluster slower and more massive. In response, gravity would pull the star cluster more tightly together, and the stars would end up moving *faster*, as predicted by the virial theorem. As the random speeds of stars in the cluster increased, we could imagine a steady progression from a nucleus in which most encounters lead to coalescence to a nucleus in which most encounters lead to disruptive collisions.

Other effects would also be important. During the coalescent phase, the rate of supernova explosions would increase as the number of massive stars did. By flinging matter out of the nucleus at high speed, numerous supernova explosions would oppose and possibly reverse the contraction of the cluster – the loss of mass would decrease the strength of gravity. The supernovae would leave behind an admixture of neutron

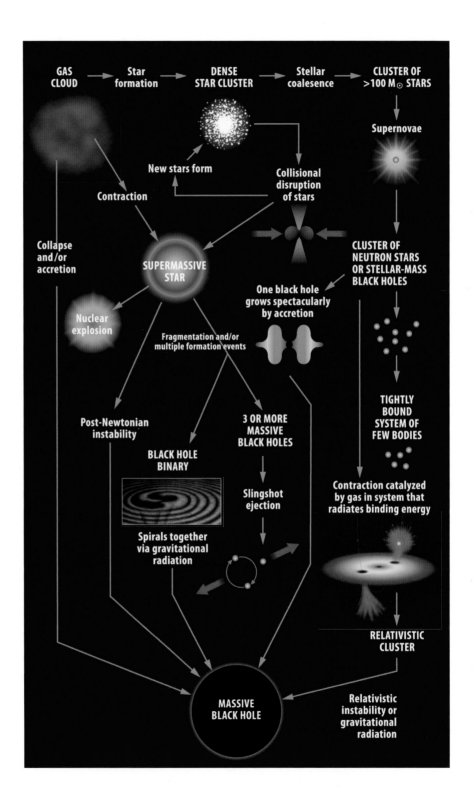

In the central part of a galaxy (the innermost few light-years) the concentration of stars can be so high that collisions are frequent. Gravity also causes unconsolidated gas to accumulate in the center; this gas could form new stars, catalyze the contraction of the existing stellar population, or collapse directly to a black hole. This diagram depicts various evolutionary tracks that might occur, and their interrelationships. The formation and/or growth of a massive black hole is the almost inevitable endpoint (literally, the "bottom line") for most evolutionary tracks. Such black holes are implicated in the most energetic phenomena in galactic nuclei, the subject of our next two chapters – in these objects we are probably witnessing episodes of vigorous black hole growth. In Chapter 8 we present evidence that massive black holes are also present in the nuclei of essentially all large, "quiet" galaxies, including the Milky Way.

stars and stellar-mass black holes, which would suffer fewer direct collisions because of their compact size but would still interact gravitationally. Weak gravitational interactions between stars that never physically touch will tend to drive the nucleus toward a "core–halo" structure with "mass segregation": stars of average mass would be ejected to larger distances, while the heavier remnants would remain in the contracting central regions (similar evolution would occur in a globular cluster, as discussed in Chapter 1). The fate of the cluster depends on the outcome of the three-way competition among the processes of direct collisions, weak gravitational encounters, and explosive mass loss. At one extreme, the coalescence process might lead to the creation of a few enormous stars that would sink to the center of the cluster, merge, and ultimately collapse to a black hole. At another, a small fraction of the initial cluster might contract to form a tightly bound stellar core in which most collisions are violently disruptive. The matter liberated by disruptive stellar collisions would be unable to escape the gravitational pull of the cluster; most of it would collect at the cluster's center, where eventual collapse to a black hole is almost inevitable.

All this time, the nucleus is interacting with its evolving galactic surroundings – it is far from a closed system. Supernova explosions or powerful stellar winds could expel much of the gaseous debris from the central region, preventing it from accumulating. This is especially likely in a galactic nucleus of modest mass and random velocity, perhaps the nucleus of a dwarf galaxy or the center of a pregalactic halo that has not yet finished merging. But in more massive or more condensed galactic nuclei, the accumulation of matter seems inevitable. Some of the gas far from the nucleus – whether released by stellar evolution in the galaxy, accreted from intergalactic space, or injected by a recent merger – is bound to drift toward the center as it loses angular momentum. Once in the nucleus, it has nowhere to go and must accumulate. What is its fate? Will it form stars, adding to the crowding in the central star cluster? Or will it collect to form an amorphous gas cloud, dragging on the orbits of passing stars and hastening their coalescence? Could it contract under its own gravity to form a single "superstar," so unstable according to theory that it would collapse to a black hole all at once? Our answers to any of these questions remain highly speculative, but we should not be surprised to find very massive black holes, of perhaps millions or even billions of solar masses, in the nuclei of many galaxies.

The Center of the Milky Way

In the face of so many complex, competing mechanisms, the only safe course is the empirical one. What do we *observe* to be going on in the

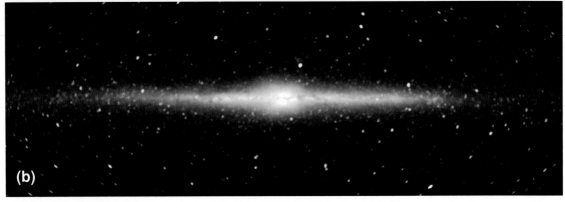

Images of the Milky Way in (a) visible light and (b) infrared. Clouds of dust and gas, which obscure our view at visible wavelengths, are transparent to infrared rays. The images are aligned with the Galactic Center in the middle.

only galactic nucleus that we can study in some detail? The nucleus of the Milky Way lies only 25 000 light-years away, compared to the more than 2 million light-years that separate us from the nucleus of Andromeda. Although optical and ultraviolet light from the Galactic Center is almost totally obscured by gas and dust, we can readily map out the region at radio, infrared, X-ray, and gamma-ray wavelengths. What we see can only be described as a mess. A dense star cluster is embedded in an extremely complex array of gaseous filaments, disks, and blobs. Large filaments with linear striations, mapped in the radio synchrotron emission of energetic electrons spiraling in a magnetic field, remind us of the prominences that erupt from the surface of the Sun. X-ray images show plumes of hot gas, along with hundreds of X-ray emitting point sources. A surprisingly large number of these are X-ray binaries – tracers of a large population of neutron stars or black holes – implying that coalescence and mass segregation may have played a role in shaping the nucleus, as theory suggests.

At the exact center of the Milky Way, there is a flaring, point-like source of radio, infrared and X-ray emission with the unromantic

Radio observations of the Galactic Center region show striated filaments of synchrotron radiation. The appearance, similar to that of a solar prominence, suggests that we are witnessing huge eruptions of the magnetic field.

An X-ray image of the Galactic Center shows plumes and bubbles of hot gas, centered on the putative black hole Sgr A*. The many point sources emitting X-rays correspond to a variety of different kinds of objects, including a surprisingly large number of X-ray binaries.

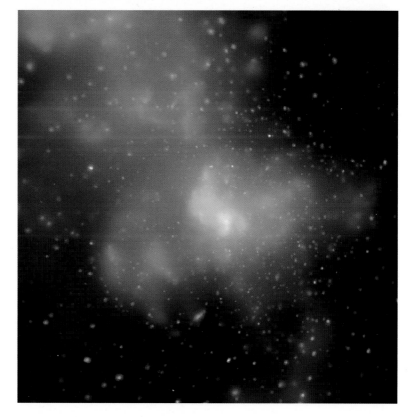

name of "Sagittarius A-star" (Sgr A*). Its compact size (it must be small enough to fit inside the Earth's orbit) and unusual radiation spectrum – which resembles the kind of spectrum that might be produced by gas swirling into a black hole at a very low rate – have long been recognized as possible signatures of a massive black hole. Another early sign was the slow motion of Sgr A* on the sky, indicating that it was very likely much more massive than a normal star. If it were only a few times as massive as a normal star, it would most likely move at a sizable fraction of the other stars' velocity. But in fact it is virtually stationary.

Measurements of stellar and gaseous motions suggested that a large mass occupied the inner light-year of our Galaxy, but in those early days (barely more than a decade ago), detractors of the black hole hypothesis pointed out that there could well be enough ordinary stars to account for all of the mass present in the domain of the central star cluster, so a black hole was not *needed* to explain the observed motions. They also worried that the energy output of Sgr A* is implausibly *low* for a very massive black hole sitting in such a dense environment of stars and gas.

In the end, the black hole proponents won: measurements of individual stars near Sgr A* clinched the case. Maps of their orbital motions – made possible by new observational techniques in the past decade – have provided the definitive evidence that Sgr A* marks the position of a black hole – or at least a "compact dark object" – of several million solar masses. The low energy output remains puzzling, though perhaps not as implausible as it once seemed. We will defer our discussion of this evidence until Chapter 8, where we will also ask how black holes can remain so "quiet" in the vast majority of galactic nuclei. In the meantime, we will focus on the relatively rare cases in which massive black holes are putting on spectacular displays of pyro-technics in the nuclei of galaxies. Some of these black holes approach a billion solar masses or more, dwarfing the black hole of our own Galactic Center. They produce the most luminous sustained phenom-enon known anywhere in the Universe – quasars – and occasionally shoot jets of ionized gas across millions of light-years.

5

QUASARS AND KIN

Not long after "spiral nebulae" and their elliptical cousins were recognized as "island universes" of stars and gas separate from the Milky Way, astronomers began to realize that something strange was going on in the centers of many galaxies. Often, there lurked intense concentrations of blue light. This light had characteristics quite unlike the radiation associated with aggregates of stars and gas, the normal components of galaxies. Its spectrum contained too much blue and ultraviolet light to come from ordinary stars, even if they were very hot. Sometimes these central sources of energy were as bright as the entire surrounding galaxy, and later it was discovered that their brightness often varied. Galaxies containing these central sources came to be known as *active galaxies*, and the central sources themselves are called *active galactic nuclei*, or AGN for short. The most extreme examples are called *quasars*. Observations made in the radio band revealed even more bizarre behavior, the narrow, fast jets of gas to which we devote the next chapter.

The discovery and interpretation of active galaxies and quasars was a convoluted process, with many false starts and wrong turns. Perhaps more than any other phenomenon discussed in this book, the discovery of the nature of AGN involved serendipity in its purest form. AGN even trump pulsars on this score, since astronomers were already looking for neutron stars in 1968 and Franco Pacini had even suggested that spinning magnetized neutron stars might radiate – although he hadn't guessed that they would pulse! *No one* had predicted or even speculated that galaxies could be "active." It required accidental discoveries by both radio and optical astronomers before it became clear that observers had found one of the most energetic phenomena in the Universe – and, as it turned out, the "smoking gun" of massive black holes in the nuclei of galaxies.

Signs of Activity

Evidence that some form of violent activity was occurring in the centers of galaxies had trickled in for nearly half a century before quasars were finally discovered in 1963. Heber D. Curtis, then Director of Lick Observatory near San Jose, California, had discovered in 1917 that the nearby galaxy M87 had a jetlike feature emanating from its nucleus. He remarked on its peculiarity but did not follow up his observation. Other nearby galaxies, such as M82, seemed to be undergoing some kind of violent disruption, although it later turned out that M82 belongs to a subclass of galaxies, called "starbursts," whose activity results not from a central black hole but rather from hyperactive star formation.

No firmer clue that something unusual was taking place in some galaxies appeared until World War II. Then Karl Seyfert, at the

Heber Curtis, of the Lick Observatory in California, discovered as early as 1917 that there was a "jet" emerging from the center of the giant elliptical galaxy M87 in the Virgo Cluster.

The jet that emerges from the nucleus of the elliptical galaxy M87 is about 6 thousand light-years long. This *Hubble Space Telescope* image shows the complex pattern of ultraviolet and visible emission from the jet as blue, in contrast to the smooth (yellow) glow of the stars. The bright "knots" could be shock waves, and the side-to-side "wiggles" might represent insta-bilities or magnetic structures. Images at radio, infrared, and X-ray wavelengths reveal strik-ingly similar features.

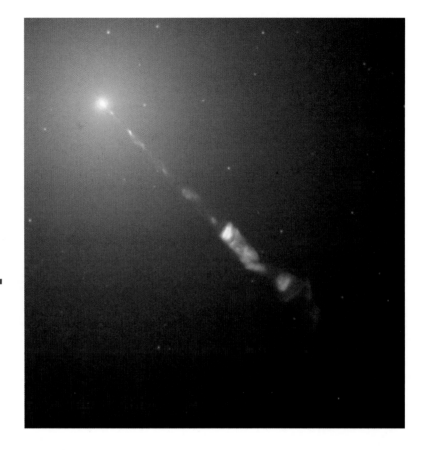

Mount Wilson Observatory in California, noted that a sizable subset of spiral galaxies – as many as a few percent – have intense blue pointlike nuclei. Spectra of what we now call "Seyfert galaxies" show strong emis-sion lines of the type produced by clouds of ionized gas. Instead of being sharp like emission lines observed in the laboratory, however, these lines are smeared over a surprisingly large range in wavelength. This smear-ing is believed to be due to the Doppler effect and indicates that the gas emitting the lines is extremely turbulent, with random velocities as large as a few percent of the speed of light – 10 to 100 times faster than the ordinary gas motions found in a galaxy. The Armenian astronomer B. E. Markarian, while making a survey similar to Seyfert's, later found that some elliptical galaxies also have intense blue nuclei.

But it was not until the radio astronomers weighed in that the astronomical community was forced to confront the possibility that some galaxies might be more than just aggregates of ordinary stars and gas. In 1954 Walter Baade and Rudolf Minkowski, of the Mount Wilson and Palomar Observatories, were trying to identify a source of radio

This image of the nearby "starburst" galaxy M82 superimposes observations made in visible light (orange and yellow-green: *Hubble Space Telescope*), infrared (red: *Spitzer Space Telescope*), and X-rays (blue: *Chandra X-ray Observatory*). New stars are forming at a furious rate near the center of M82. Frequent supernova explosions create a hot wind that escapes through two channels along the rotation axis of the nearly edge-on galactic disk. The orange filaments, glowing in the Hα spectral line of hydrogen, outline pockets of 10 000-degree gas on the fringes of the wind; the interior, hotter still, is observable only in X-rays.

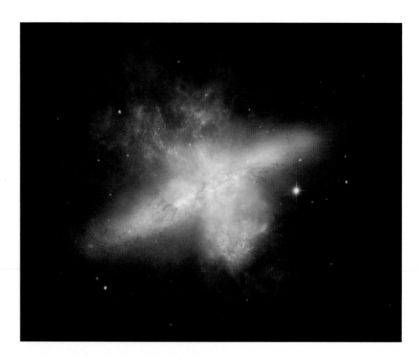

waves called Cygnus A, the second "brightest" object in the radio sky. At that time astronomers knew only that something was emitting a lot of radio waves, but they didn't know what, or even how distant, the object was. Baade and Minkowski's breakthrough was to associate Cygnus A's position on the sky with a very faint galaxy having a redshift of 0.05, thus placing it about 300 times farther away than the Andromeda Galaxy. This distant galaxy was evidently the source of the radio emission. If a galaxy that produced such faint optical signals could produce such loud radio signals, then the hope was raised that some peculiar galaxies might be detectable by radio techniques even if they were so far away that the combined light from 10^{11} stars failed to register optically.

The faint optical image of Cygnus A showed a seemingly disturbed structure made of two distinct parts, which Baade and Minkowski interpreted as the signature of two colliding galaxies. Shortly thereafter, Roger Jennison and M. K. Das Gupta, of the Nuffield Radio Astronomy Laboratories at Jodrell Bank near Manchester, England, showed that the radio emission from Cygnus A was not coming from the galaxy at all, but from two giant patches placed symmetrically about the galaxy on the sky, a quarter million light-years to either side! These radio patches are thought to have a three-dimensional structure like a dumbbell, hence they are often called "lobes." This double radio structure is now known to be characteristic of the strongest sources of radio waves, and we shall have

If an array of randomly moving gas clouds emits light in a spectral line, the receding clouds will appear reddened relative to the approaching clouds. The light from the whole system, when spread out into a spectrum, yields a line that is much broader than that from an individual cloud. The overall width is determined by the spread in cloud velocities; the shape of the line indicates the relative number of clouds with different velocities.

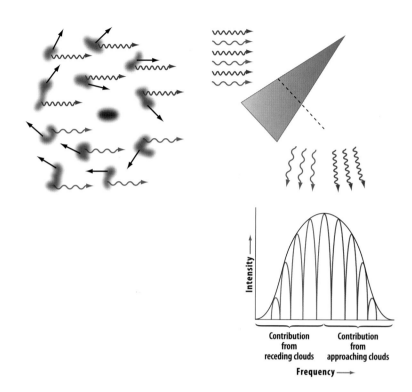

Contribution from receding clouds | Contribution from approaching clouds

Frequency ⟶

Intensity ⟶

much more to say about its interpretation in the next chapter. Jennison and Das Gupta's discovery cast doubt on the collision hypothesis (and the double *optical* structure proved to be an illusion created by an obscuring lane of dust), but left the mystery of the energy source for the radio emission more urgent than ever.

It was already recognized in the 1950s that the radio emission from cosmic sources was synchrotron radiation emitted by electrons gyrating in magnetic fields at nearly the speed of light, just as in the Crab Nebula. To produce the observed amount of synchrotron emission, the lobes would have to contain energy in the forms of both fast-moving electrons and magnetic fields. But how much energy? In celebrated papers published in 1956 and 1959, Geoffrey Burbidge showed that the total energy content of the radio lobes of an extended source such as Cygnus A – the combination of magnetic fields and fast particles – must exceed the amount that would be released by the *complete annihilation* (via $E = mc^2$) of a million solar masses of material.

Burbidge's result was the first indication that galactic nuclei can release energy on scales that vastly exceed even that of a supernova explosion, and that somehow this energy is channeled into the form of particles moving very close to the speed of light – in other words, a

Walter Baade and Rudolf Minkowski identified the radio source Cygnus A – the second brightest source in the sky (not counting the Sun) – with this faint galaxy 300 times farther away than the Andromeda Galaxy. The dust lane running across the galaxy gave it a double appearance, which misled Baade and Minkowski into proposing that two galaxies were colliding. The radio output is actually energized by a concentrated power source in the galactic nucleus.

"relativistic" plasma – and magnetic fields. Where did this energy come from? The great Armenian astrophysicist Victor Ambartsumian (who, as director of Byurakan Observatory in Soviet Armenia, had sponsored Markarian's survey) speculated that these "exploding galaxies" were the places where matter was being created, as required by the then-popular "steady state" theory of cosmology propounded by Hermann Bondi, Fred Hoyle, and Thomas Gold. (According to this theory, the gross structure of the Universe was unchanging despite the cosmic expansion, because matter was continuously being created. Despite its philosophical appeal, this theory has been discarded because of compelling evidence that the expansion of the observable Universe began a fixed time in the past, and that distant galaxies at high redshifts, whose light set out when the Universe was younger, look systematically different from their nearby counterparts.) Although Ambartsumian's conjecture was forgotten when the steady state theory fell into disrepute, other speculations abounded as well – for example, it was suggested that galactic nuclei were giant matter–antimatter annihilation reactors. No one yet suspected black holes.

The major contribution of optical astronomy to this story came in 1963. Cyril Hazard, M. B. Mackie, and John Shimmins, using the recently completed Parkes Radio Telescope in Australia, managed to pin down the position of a particular radio source, 3C 273,

The Very Large Array in New Mexico reveals the double-lobed radio structure of Cygnus A in rich detail. The energy produced in the nucleus is carried to the lobes by jets, the traces of which are visible as thin lines. "Hot spots" at the end of each lobe are thought to be shock fronts where the jets run into the intergalactic gas. The source extends about half a million light-years end to end; the galaxy photographed in visible light is less than a tenth this size.

to a precision of a few arcseconds, by noting the time at which it was occulted by the Moon. They had to make special adjustments to the telescope to observe something as far north – and consequently as close to the horizon – as 3C 273. ("3C" denotes the third Cambridge catalog of radio sources – the source had been discovered by means of a radio telescope in Cambridge, England.) Fixing the position of a radio source was a real coup at that time, since radio telescopes had much poorer angular resolutions than they do today. The ability of any kind of telescope to pinpoint a location on the sky is fundamentally limited by the ratio of the aperture of the telescope to the wavelength of the radiation being observed. Radio waves have wavelengths of a few centimeters, so to obtain a resolution of 1 arcsecond (5×10^{-6} radian) would require a telescope several kilometers across! In photographs taken in visible light, the only object coincident with the radio position of 3C 273 on the sky was a 13th-magnitude starlike object. Although an object with this magnitude is about a thousand times fainter than the faintest stars visible to the unaided eye, it is not especially faint by the standards of modern astronomy. Ground-based optical telescopes equipped with electronic sensors routinely observe stars and galaxies that are more than 10 000 times fainter than 3C 273.

Once astronomers had located an object that seemed to be the radio source 3C 273, Maarten Schmidt used the 200-inch Hale telescope on Mt. Palomar to take a spectrum of it. At first the spectrum seemed like gibberish – none of the spectral lines seemed to be at the wavelengths produced by known atomic elements. However, Schmidt

Maarten Schmidt was the first to recognize that quasars are ultra-luminous distant objects. After Cyril Hazard and his colleagues in Australia had pinned down the position of the radio source 3C 273, Schmidt took a spectrum of the starlike object in that location. The unusual features of the spectrum could be understood only if there was a 16 percent redshift, implying an enormous distance. Schmidt has, over the past 40 years, continued to discover and study distant quasars. He has been a leader in analyzing how the quasar population depends on redshift (or cosmic epoch), as described in Chapter 9.

eventually noticed that the *relative* wavelengths of the lines fit the pattern formed by glowing hydrogen, but redshifted by 16 percent. At first he was baffled, since an object at that redshift would be nearly 2 billion light-years away, if it were receding from us according to Hubble's law of cosmic expansion. At that distance, even an entire galaxy would have been much fainter than 3C 273. Yet Schmidt finally convinced himself that 3C 273 was actually at a cosmological redshift of 0.16, and was therefore by far the most luminous object ever detected. Once the intellectual block had been removed, it was quickly realized that the hitherto mysterious spectra of other objects of the same class also displayed high redshifts. A check of photographic archives, kept at Harvard College Observatory since the late nineteenth century, revealed that 3C 273 fluctuated in brightness on a timescale as short as a month. Thus, there seemed to be a class of objects that looked like ordinary stars on photographic plates, but showed spectral emission lines with high redshifts. These objects, dubbed quasars or QSOs (for quasi-stellar objects), could outshine an entire galaxy by more than a factor of 100!

Part of the reason that quasars were not recognized sooner was that they superficially resemble ordinary stars. They surrendered their extraordinary secret only upon careful study with spectrographs, performed after their radio emission had been detected by chance. Quasars have just the properties envisaged by John Herschel in 1820, in a draft of his address to the newly founded Royal Astronomical Society: "Yet it is possible that some bodies, of a nature altogether new, and whose discovery may tend in future to disclose the most important secrets in the system of the Universe, may be concealed under the appearance of very minute single stars no way distinguishable from others of a less interesting character, but by the test of careful and often repeated observations."

Quasars are now known to be just very luminous versions of the same blue nuclei that Seyfert observed in the centers of nearby spirals. The reason that they first appeared to us as isolated stars is that they are so distant, and so luminous compared to the surrounding galaxy, as to render it nearly invisible. It took the Hubble Space Telescope, in the mid-1990s, to routinely detect the host galaxies of quasars. That most of the early quasars discovered had strong radio emission is not surprising, since optical astronomers used the radio emission as a tipoff to decide which objects to study. However, we now know that quasars with strong radio emission are the exception rather than the rule. In nine out of ten cases, the intense pointlike source with broad (that is, Doppler-smeared) emission lines is unaccompanied by strong radio emission.

Whereas quasars and Seyfert galaxies are especially luminous at optical, ultraviolet, and X-ray wavelengths, other galaxies show anomalous activity of a very different sort. *Radio galaxies,* of which Cygnus A

Quasar spectra of different red-shifts, z, show a variety of prominent lines. The redshift stretches the wavelengths by a factor (1 + z) between emission and detection. In quasars with redshifts of 2 or more, the strongest line observed at visible wavelengths is the Lyman α line of hydrogen, which is emitted in the far ultraviolet. Since far-ultraviolet radiation cannot penetrate the Earth's atmosphere, this line is never observed from the ground except when it is redshifted.

and M87 (sometimes called Virgo A by radio astronomers, since it is the brightest radio source in the constellation Virgo) are archetypes, somehow manage to channel the bulk of their energy outputs into the radio band, producing large diffuse patches of radio emission on the sky. Astronomers love to invent classification schemes for new phenomena, and the many different varieties of "active galaxy" have provided a fertile outlet. The major classes of Seyferts, quasars, and radio galaxies are all subdivided in a variety of ways, often according to overlapping observational criteria. Unfortunately, observers were often more perceptive in distinguishing the different classes than they were imaginative in naming them. As a result, we routinely use the nondescriptive terminology of Type 1 and Type 2 for Seyfert galaxies (and sometimes, now, for quasars), "Fanaroff–Riley types" 1 and 2 for radio galaxies, and other such nomenclature. (This seems to be a general problem in astronomy: supernovae, for example, are classified as Type 1, Type 1a, Type 1a-peculiar, Type 1b, Type 2, etc.; stellar populations are Pop I and Pop II, etc.) Additional subcategories were invented to represent transitional objects. For example, highly variable, polarized AGN with very weak or undetectable emission lines are placed in a special category called BL Lac objects, so-named because the prototype had been misclassified as a variable star, BL Lacertae.

All the varieties of quasars, Seyferts, and radio galaxies are considered to be members of the observationally heterogeneous class of objects known collectively as active galactic nuclei. But how much do these objects really have in common? Although spiral and elliptical galaxies have very different appearances, stellar populations, and internal dynamics, they share fundamental structural similarities as self-gravitating systems of stars and gas. Therefore, lumping them together as *galaxies* amounts to a powerful generalization that has enabled astrophysicists to uncover principles that unify the types of structure found in the Universe. Does the term "active galactic nucleus" similarly reflect a unity among these seemingly diverse objects, at a deep level?

One obvious link is that all forms of activity in these galaxies originate in the nucleus. (The connection is obvious for quasars and Seyferts, because of their bright compact nuclei; we shall see in Chapter 6 how the connection has been established for radio galaxies.) But one can go further. Generic similarities among the different forms of activity suggest that they all represent different manifestations of the same kind of phenomenon. One might group these generic properties into four categories, as follows.

- *AGN can emit energy at an enormous rate.* The power outputs of the most luminous quasars and the most powerful radio galaxies can exceed the total luminosity of *all* the stars in a large galaxy by more than a factor of 100. Episodes of activity can apparently last for many millions of years. AGN are the most powerful sustained phenomena known in the Universe.

- *AGN are extremely compact.* We have already noted that the quasar 3C 273 changes in brightness in the course of a month. Rapid variability – on timescales as short as a few hours in some cases – is a generic feature of AGN. This variability puts an upper limit on the size of an AGN. If an astronomical body routinely varies in brightness by a factor of 2 over a certain time interval, one can generally infer that it must be smaller than the distance light can travel during the same interval. No signal can travel faster than the speed of light to tell different parts of the object to vary in step.

  Consider the following analogy. The final game of the Intergalactic World Cup is being held on Earth, with the Milky Way team playing the Andromeda Galaxy. Four planets – Earth, Mars, Jupiter, and Saturn – all happen to be lined up on one side of the Sun, and lie in the plane of the sky as viewed from Andromeda. To take advantage of this unusual arrangement, an executive from the ABC (Andromeda Broadcasting Corporation) Network proposes the following stunt: the four planets will be wired so that when a switch is flipped on Earth, they light up to spell "GOAL." Each planet is to display one letter – "G" on Earth, "O" on Mars, "A" on Jupiter, and "L" on Saturn. The idea is that Andromedan sports fans with powerful telescopes can then see the Solar System light up as soon as a goal is announced over the ABC Network. Unfortunately, the plan doesn't work very well. True, as soon as they hear about the goal via a direct radio link from the Earth, they see the Earth light up with a "G." But 4 minutes pass before the "O" on Mars switches on, another 31 for the "A" on Jupiter, and an additional 36 for the "L" on Saturn. Altogether, it has taken one hour 11 minutes for the sign to turn on, by which time the game is over. The reason, of course, is that it takes light 71 minutes to travel from the Earth to Saturn. There is no way, even in theory, that Saturn could know about the goal less than 71 minutes after it occurred. Likewise, if a quasar were as big as the distance from the Earth to Saturn, it could not change in brightness significantly in less than about 71 minutes. In fact, quasars seldom vary this rapidly, but the observed rate of variability does imply that they cannot be too much larger than the Solar System – a maximum size that is tiny by galactic standards.

- *AGN radiation is unlike that normally produced by stars or gas.* Although AGN radiate most strongly in one part of the spectrum or another, they always produce substantial radiation over a

The interplanetary "GOAL" sign at the Intergalactic World Cup. Since it takes 71 minutes for signals from Earth to reach Saturn, traveling at the speed of light, the Andromedans do not see all the letters of the sign light up simultaneously. Reversing the argument, we can estimate the maximum possible size of a variable radiation source – such as a quasar – by assuming that the typical variation timescale is the light travel time across the source.

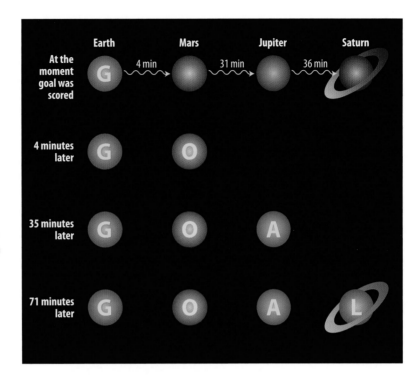

very wide range of wavelengths. Some AGN emit comparable amounts of power in the radio, infrared, optical, ultraviolet, X-ray, and gamma-ray portions of the spectrum, a factor of more than 10^{10} in wavelength! Stars and gas, by contrast, tend to emit the bulk of their radiation over a relatively narrow band of wavelengths. That characteristic band of wavelengths is inversely correlated with the temperature of the radiating material, according to the so-called blackbody laws describing radiation from solids, liquids, and dense gases. Thus a very hot body will radiate mostly in a band of very short wavelength, in the X-ray or gamma-ray portion of the spectrum. Radiation characterized by a well-defined peak in the spectrum is called "thermal," because it resembles that produced by a radiating solid of fixed temperature, such as the filament in a light bulb or the heating element in an electric stove. AGN have "nonthermal" spectra, because the power is so evenly distributed in wavelength that it is impossible to associate a temperature with it.

• *AGN contain gas moving at extremely high speeds, sometimes approaching the speed of light.* We can deduce the velocities of gas clouds in quasars and other types of AGN by measuring the Doppler shifts of spectral emission lines. The velocities measured by this method typically range up to 10 percent of

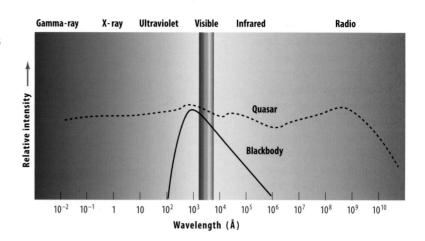

Gamma-ray   X-ray   Ultraviolet   Visible   Infrared   Radio

A "black body" absorbs all the energy that falls on it and radiates at a well-defined temperature. The surfaces of stars, for instance, emit approximately "blackbody" radiation whose spectrum has a characteristic shape and peaks at shorter wavelengths (bluer colors) for hotter bodies. A similar plot for a quasar, however, shows that energy is distributed over a much broader range of wavelengths. This is because several regions with different temeratures contribute to the total spectrum, and also because some of the emission comes from processes (such as synchrotron radiation) that do not have a well-defined temperature. The solid curve indicates what shape a quasar spectrum would have if the radiated power came from immediately around the black hole and had a spectrum like that of a star.

the speed of light. As we discuss in Chapter 6, the velocities of jets produced by radio galaxies are usually estimated by more indirect methods; nevertheless, the evidence that they contain gas flowing away from the nucleus at very nearly the speed of light is compelling.

The Black Hole Paradigm

What can we deduce from these shared characteristics of AGN? We shall argue below that all these features can be explained most simply as manifestations of massive black holes at the centers of galaxies. The crux of the argument is contained in the first two characteristics, which together are so extraordinary that they bear restating: AGN are objects that can produce a galaxy's worth of power (or more) in a region the size of the Solar System. Since by definition the gravitational field near a black hole can accelerate gas to near the speed of light, it is not surprising that the fourth characteristic can also be accommodated readily within the black hole paradigm. However, the details of how high-speed gas streams are produced seem to be rather complicated; we shall explore some aspects of their creation in Chapter 6. Nonthermal radiation spectra (the third characteristic) are also consistent with the black hole paradigm, but here the details are more debatable; we reserve further discussion on this topic until the end of this chapter.

The reasoning leading to the black hole paradigm is surprisingly simple. Since AGN vary rapidly in brightness, we can place *upper limits* on their sizes *(R)*. At the same time, we can place *lower limits* on the amount of mass contained in the region where the luminosity is produced *(M)*. This means that we can derive a *lower limit* on the ratio of mass to size, M/R – let's call this the "mass compactness." The mass compactness turns out to be a direct measure of the strength of the gravitational field in an object, and for a black hole it attains a unique maximum value that depends only on constants of nature – c^2/G, where c is the speed of light and G is Newton's universal constant of gravitation. Thus, by comparing M/R (estimated from observations) to c^2/G (predicted by theory), we can tell how "close" an object is to being a black hole. Needless to say, general relativity predicts that no object can have a mass compactness that exceeds c^2/G!

We estimate size as follows. If the luminosity changes significantly (say, by a factor of 2) during a typical time interval t, then it is probably safe to assume that the source cannot be much bigger than the distance light can travel during the same amount of time – that is, R is less than $c \times t$, where once again c is the speed of light. (There is an exception to this rule, when the source is moving toward the observer at nearly the speed of light. This effect is crucially important if one wants to estimate the properties of jets, as we shall see in Chapter 6, but is not believed to be very important for optical emission from quasars and Seyfert galaxies.)

Mass estimates can be made in several ways. One method utilizes Einstein's law of mass-energy equivalence, $E = mc^2$. If we can find out the total energy radiated by an AGN over its lifetime, we can use this relation to calculate the amount of mass converted into energy. No plausible process can convert matter into energy with perfect efficiency; the best one can hope for is an efficiency ranging from 1 percent (for nuclear reactions) to about 40 percent (for matter spiraling into a spinning black hole). Matter–antimatter annihilation, which does yield 100 percent conversion, can be ruled out as the main power supply by other arguments. Thus, at least half the "fuel" that enters the "central furnace" of an AGN retains its form as matter. What happens to this unspent fuel? Given the inexorable pull of gravity, the better part of it remains behind, accumulating in the center of the galactic nucleus. Thus, given the lifetime energy output of an AGN, we can estimate a *minimum* mass for its furnace.

Astronomers have devised at least two techniques for estimating the lifetime energy output of an AGN. If the active nucleus contains a powerful radio source, we can use Burbidge's method to calculate the minimum energy that would be required to produce the observed synchrotron radiation. This method works because the radio lobes store a large fraction of the *cumulative* energy output of the AGN. Alternatively, if we can guess the *age* of a source, we can estimate the lifetime output by multiplying the current luminosity by the age. Nobody has found a way to measure the age of an AGN with any reliability, but we can use statistical arguments to place a lower limit on the average lifetime of an AGN. That limit is deduced from the fact that about 1 percent of all galaxies are observed to be active at any one time. If the typical duration of activity were shorter than about 100 million years, then there wouldn't

For more than 25 years, Fred Hoyle was perhaps the most versatile and original figure in the world of astrophysics. He pioneered the study of how the chemical elements are synthesized inside stars, co-invented the "steady state" cosmology, and contributed many ideas on stellar and galactic structure. In the early 1960s, he was one of the first to emphasize the importance of gravitational energy and supermassive objects in galactic nuclei.

be enough galaxies to account for all the active ones we see (since 100 million years is about 1 percent of the age of the Universe). With an estimate of the age of an AGN in hand, we can perform the sequence of steps needed to calculate a minimum mass: age times luminosity gives total lifetime energy output; assuming an energy conversion efficiency of (say) 10 percent gives the total mass "processed" through the nucleus, of which (say) at least half remains behind in the AGN.

Using a different argument, we can obtain a lower limit on the mass directly from the luminosity of the AGN. Radiation exerts a force, called radiation pressure, that is negligible in the laboratory but must be reckoned with in an object as powerful as a quasar. If the ratio of luminosity to mass in the AGN is too great, the force exerted by the radiation is great enough to blow the matter out of the gravitational field. The limiting luminosity for a given mass is called the "Eddington limit" (after the British theorist who appeared in Chapter 2 as Chandrasekhar's nemesis). Since AGN seem to persist for a considerable time without dispersal, we can guess that their luminosities do not exceed the Eddington limits for their respective masses, and this assumption leads to a lower limit on the mass.

How do the numbers work out? Fortunately, the masses estimated by different methods seem to be in agreement with one another. A typical luminous quasar, or powerful radio galaxy, must have a central mass on the order of 100 million to a few billion Suns. Less luminous Seyfert galaxies can get by with somewhat lower masses. Having obtained a size and a mass, one can now compute the mass compactness M/R. But the variability of AGN is seldom fast enough, and thus the value of R small enough, to clinch the black hole identification without further argumentation. The best we can do by measuring variability is to conclude that M/R must exceed $0.001 c^2/G$ in typical AGN. This is 100 to 1000 times more compact than a normal star and much too compact to be a star cluster. For a few objects, such as M87 and the Seyfert galaxy MCG–6-30-15, other measurement techniques lead to higher estimates of M/R – we discuss these special cases in Chapters 6 and 10, respectively. We should remember that in every case, the measured value for M/R represents the minimum possible compactness – for instance, we have no reason to expect that the luminosity of an AGN fluctuates at the maximum rate consistent with its size and the speed of light. But lower limits on M/R, by themselves, do not offer ironclad proof in favor of black holes.

Yet we believe that massive black holes lie at the centers of active galactic nuclei. Why? Partly by a process of elimination. Just as we saw in Chapter 3 that compact stellar remnants more massive than about 3 solar masses can hardly be anything but black holes, so have astrophysicists failed to imagine any realistic alternative to massive black holes in AGN. From time to time alternatives have been proposed,

Geoffrey Burbidge recognized, as early as 1956, the colossal energy requirements of extragalactic radio sources, and played a leading part in the debates, during the 1960s and thereafter, about the nature of quasars. He maintained a collaboration with Hoyle that lasted more than 40 years, covering nucleosynthesis, galactic nuclei, and cosmology.

but none has survived close scrutiny. As we saw in Chapter 4, a variety of runaway processes are expected to occur in galactic nuclei composed of ordinary stars and gas. Could these processes result in highly compact and luminous nuclei of some form other than black holes? We are not yet able to predict exactly where such runaway processes will lead. Not only would the stars interact with one another gravitationally, but also stellar collisions and interactions with gas would be important. Computer models are starting to grapple with these complexities, but we still do not know whether a dense cluster of ordinary stars would dissolve into a single supermassive star or into a cluster of neutron stars and stellar-mass black holes that would then evolve further via gravitational interactions. Some of the collisions that hasten the collapse of an ordinary star cluster can be avoided once the stars themselves are compact. But even these superdense star clusters would be subject to instabilities (predicted by general relativity) that would lead to their prompt collapse to a single massive black hole. The prospects for avoiding the formation of a black hole do not look promising.

Could a "supermassive star" be the long-lived "central engine" of an AGN? Probably not. Any star much more massive than 100 solar masses would be supported against gravity by the force of radiation pressure and would have a luminosity very nearly equal to the Eddington limit corresponding to its mass. But this equilibrium is precarious. Such stars are barely stable according to the Newtonian theory of gravity, and the corrections required by general relativity, even when small, are destabilizing. Even if stabilized by rotation or a strong magnetic field, such a "star" would be unlikely to survive more than a few million years before collapsing, and could not represent more than a brief transient phase on the way to a massive black hole.

Although many of the arguments presented above seem self-evident in retrospect, adoption of the black hole paradigm did not come quickly or easily. One group of prominent researchers, among them Fred Hoyle and Geoffrey Burbidge, was quick to recognize that only a process involving gravitational collapse could meet the prodigious energy requirements of radio lobes. They considered several candidates – including supermassive stars and dense star clusters – and noted the shortcomings of each. But they despaired of finding a sensible mechanism to explain both the luminosity and compactness of quasars, and for years afterwards insisted that these objects must be relatively nearby (and therefore, less luminous than they seemed), large redshifts notwithstanding. It was not until 1969 – 5 years after Edwin Salpeter and Yakov Zel'dovich had pointed out the high efficiency of accretion onto a black hole (Chapter 3) – that the British astrophysicist Donald Lynden-Bell proposed that quasars are powered by accretion disks around black holes as massive as 100 million suns or more. Just as X-ray binaries are

Donald Lynden-Bell of Cambridge University has nearly 50 years of distinguished research behind him, covering a range of topics in galaxies and cosmology. But the single contribution of which he is (rightly) most proud is a paper written in 1969, in which he conjectured that massive black holes might lurk in the centers of ordinary galaxies, and discussed the physical properties of the accretion disks that would encircle them.

fueled by the transfer of gas from the companion star onto the compact remnant, the massive black holes in AGN would be fueled through the capture of gas from the surrounding galaxy, or even by the swallowing of entire stars. This captured debris would swirl downward into the intense gravitational field, reaching nearly the speed of light before it is swallowed. The gravitational energy thus liberated would provide the power for both the luminosity and the outflowing gaseous jets that characterize AGN. Lynden-Bell also predicted that the modern-day galaxies that had once hosted quasars should still contain their black hole relics. Evidence for these quiescent black holes now abounds (Chapter 8), providing perhaps the strongest vindication of the black hole paradigm.

Modeling Quasars

It is one thing to accept that supermassive black holes interacting with local matter cause the phenomena of active galactic nuclei, and quite another to understand how these interactions lead to AGN phenomena *in detail*. We are still far from a detailed model for AGN, but certain pieces of the puzzle seem to be falling into place.

Massive black holes can generate a high luminosity in two quite distinct ways: straightforwardly by accreting mass, as in the X-ray binaries discussed earlier; or via electromagnetic processes that tap the spin energy of the black hole. In the latter case the black hole behaves like a flywheel, and the surrounding matter behaves like a "brake" that extracts energy by slowing down the spin. The flywheel mechanism may be key to the production of jets, as we shall see in Chapter 6, but there is no reason why both mechanisms couldn't operate simultaneously. The properties of an AGN must depend, among other things, on the relative contributions of these two mechanisms, which in turn depend on the accretion rate, the magnetic fields, and the spin of the hole. The observed properties of AGN must also depend on other parameters, such as the mass of the black hole and the orientation of the accretion disk with respect to our line of sight. Ideally, one would like a unified model that explains the multifarious types of AGN in the same way that theories explain the correlations among various properties of stars.

A fairly close analogy can be drawn between accretion in X-ray binaries and in AGN. Mass transferred from the binary companion supplies the fuel in an X-ray binary; in the case of an AGN, it is uncertain whether the supply of accreted matter comes from normal interstellar gas in the surrounding galaxy, driven into the nucleus during a close encounter or merger with another galaxy, or (particularly in the less luminous cases) gas released by stellar winds or via the disruption of individual stars passing too close to the hole. In any of these cases most of the gas is likely to have enough angular momentum that it cannot fall in directly, but rather must form an accretion disk. The theory of accretion disks makes specific predictions about the radiation spectrum that should be emitted by a disk with a given accretion rate, around a black hole of given mass. Whereas the spectrum of an X-ray binary is predicted to peak, as observed, at X-ray wavelengths, the accretion disks in AGN are predicted to peak at ultraviolet wavelengths. How does this prediction square with observations? Much of the ultraviolet wavelength band is difficult to observe directly because it is absorbed by intervening clouds of dust and gas. Still, it appears that the spectra of quasars do indeed exhibit a "big blue bump," which could be the radiation emitted by an accretion disk.

But an accretion disk alone cannot explain all the features of AGN spectra. This is hardly surprising, given that galactic nuclei display activity in many forms, in all wavebands from radio to gamma rays. Unfortunately for us theorists who prefer simple all-encompassing explanations for phenomena, it seems that many different physical processes contribute to the emission from AGN. The infrared radiation probably comes from dust, which has been heated to approximately 1000 degrees Celsius by absorbing some of the ultraviolet radiation produced close to the black hole. The X-rays may come from a hot, turbulent corona that forms a kind of "atmosphere" above the disk, much as the solar corona produces X-rays from a region that extends well beyond the cooler chromosphere. Computer simulations are beginning to show how such a corona might form, and indicate that it might be responsible for dissipating a large fraction of the energy liberated by accretion. Radio and gamma-ray emissions probably come from jets of gas that are accelerated parallel to the disk's rotation axis. A theme that has come to dominate many aspects of AGN modeling is that of "reprocessing." Radiation emitted in one waveband, say X-rays, can be intercepted by the disk, for example, and re-emitted in a modified form. Thus, X-rays can be reprocessed into ultraviolet radiation, and ultraviolet into optical and infrared. These complex transformations of energy from one form to another make it much more difficult to track down the "primary" sources of radiation.

Another prominent theme in AGN modeling is that of "unification." This is the idea that certain observational classifications of AGN phenomena depend only on the observer's orientation relative to the axis of the accretion disk. One successful unification scheme unites Type 1 and Type 2 AGN, according to whether the viewer has a clear line of sight to the black hole (Type 1), or is prevented from seeing the innermost region by an opaque torus of dusty gas that surrounds the thin, hot accretion disk (Type 2). In the latter case, the observer does

Because the inner parts of an accretion disk move faster than the outer parts, lines of magnetic force get sheared and stretched. The magnetic field strengthens until it forces its way out of the disk plane. The resultant loops of magnetic field can flail around and dissipate energy, rather like the magnetic arcs sometimes seen on the Sun (inset). Magnetic fields may also provide the main friction (or viscosity) that removes angular momentum and allows the material in a disk to swirl inward toward the black hole.

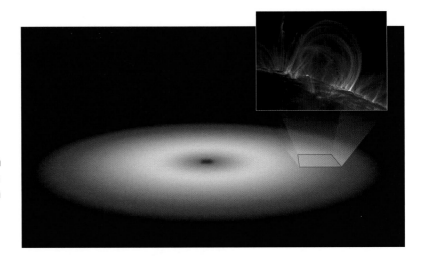

not see the ultraviolet source directly, or even the Doppler-smeared emission lines from clouds of gas directly above the disk – but *reflections* of these features off more distant gas clouds are often visible.

A second unification scheme applies to the approximately 10 percent of AGN that have powerful jets and are therefore classified as "radio-loud." As we shall see in Chapter 6, the radiation emitted by these jets appears much brighter when viewed along the axis than from the side – a consequence of the jets' speed being close to the speed of light. An observer looking down the jet sees a violently fluctuating source of radiation called a "blazar," which can be so intense that it drowns out the radiation from the accretion disk and line-emitting clouds (in which case the object would be classified as a BL Lac). From a viewing direction away from the axis, the jet would appear much more benign and the AGN would be classified as a radio galaxy or radio-loud quasar.

The success of classification schemes based on viewing direction reassures us that central engines may be similar in all highly active galactic nuclei. Accretion power and "flywheel" power could both to various degrees contribute to an AGN's primary power output. That output would then be reprocessed in a variety of ways, depending on details of the galactic environment. But many basic problems remain unresolved. For instance, we do not understand why a minority of AGN produce powerful, tightly collimated jets while the remainder do not. Nor are the unification schemes as clearcut as we would like. While all Type 2 AGN apparently possess opaque "doughnuts" of dusty gas, some Type 1 AGN apparently do not – in these objects we could see into the center no matter what our viewpoint.

Schematic diagram of an active galactic nucleus (not drawn to scale), illustrating the main structural elements, sources of radiation, and geometric origin of "unification." Energy comes out of the innermost regions in the form of ultraviolet and X-ray radiation, and (in radio-loud AGN) high-speed jets of gas. Surrounding the inner, thin accretion disk is a thick dusty torus, which absorbs radiation from the hotter inner regions and re-emits the energy at infrared wavelengths. Radiation could also be reprocessed by gas clouds or streamers distinct from the disk. Viewers whose line of sight to the inner disk is blocked by the torus see a "Type 2" AGN, while those able to look through the "doughnut hole" see a Type 1. A similar unification scheme applies to radio-loud AGN: viewers looking along the jet see a violently variable "blazar" source (a BL Lac or similar object) while those with an oblique line of sight see a radio galaxy or quasar.

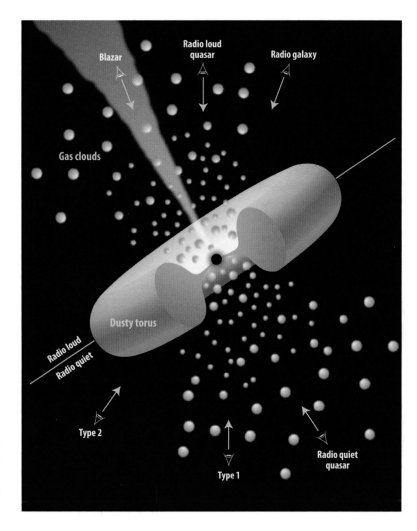

Conditions around black holes are extreme, but the relevant physics is known, and the key problem is at least well posed: given a black hole with a particular mass and spin, in a given environment, how much power is derived from accretion, how much extracted from the hole's spin, and in what forms do these respective luminosities emerge? Such calculations play the same part in the modeling of AGN that nuclear physics plays in the modeling of stellar structure and evolution. However, the confrontation of theoretical models with observations, indirect even for stars, is admittedly much more ambiguous for AGN. In stars the energy percolates to the observable surface in a relatively steady and well-understood way; in AGN, on the other hand, it is reprocessed into all parts of the electromagnetic spectrum over a huge range of spatial scales, in a fashion that depends in a

Like a searchlight beam piercing a misty sky, the active nucleus of the nearby Seyfert galaxy NGC 1068 lights up surrounding wisps of dust and gas. Hidden from our direct view (at lower left) inside a doughnut-shaped ring of opaque gas clouds, the intense ultraviolet radiation from the nucleus can escape only through the doughnut's opening. This *Hubble Space Telescope* image shows the cone of interstellar gas illuminated and ionized by the escaping rays. If we could look along the axis of the doughnut, we would be able to see the unattenuated radiation from the "central engine."

poorly understood way on the particular environment offered by the host galaxy.

What does the future hold for quasar research? If the past is any guide, the way is unlikely to be paved by theorists. In complex phenomena like AGN, too many processes *could* occur, and picking the likely winners in advance of observations has proven to be too difficult. In fact, progress in understanding quasars and other distant phenomena has mainly been driven by the advent of new techniques, new instruments, and engineering innovations. This is hardly a break with astronomical tradition – in some ways, astronomy has always been "big science." It was the first science to be professionalized and to use elaborate equipment (even though there has always been a strong amateur interest). One can see this as far back as the late eighteenth century by contrasting Herschel's telescope, a massive and elaborate construction, with the small-scale equipment that would have been used by his famous contemporaries Lavoisier (a chemist) and Cavendish (a physicist), either of whose "laboratories" would have fit on a kitchen table.

In the mid-nineteenth century, the principal impetus to astronomy was provided by the development of the spectroscope and of photography, at a time when the main theoretical tool was still Newtonian celestial dynamics. Only thereafter did people come to realize that it might indeed be feasible to study the constitutions of celestial objects.

Hubble's great book of 1936, *The Realm of the Nebulae*, concludes with these words: "With increasing distance our knowledge fades and fades rapidly. Eventually we reach the dim boundary, the utmost limits of our telescope. There we measure shadows, and we search among ghostly errors of measurement for landmarks that are scarcely more substantial. The search will continue. Not until the empirical resources are exhausted need we pass on to the dreamy realm of speculation."

The bright and beautiful photographs of galaxies seen in popular books give a misleading impression. Galaxies are barely detectable above the telltale glow of the night sky, even at the darkest sites on Earth or from space, and only long exposures reveal them at all clearly. Our view of the extragalactic Universe is still a sadly dim and blurred one. But the search continues, as more powerful telescopes and detectors are deployed. Orbiting observatories sensitive to gamma-ray, X-ray, ultraviolet, and infrared radiation, and radio telescopes on the ground, have all contributed fundamentally to our understanding of AGN phenomena.

The pace of technological breakthrough has often far outstripped the ingenuity of the theorists and phenomenologists who puzzle over the interpretation of data. Some of us have wasted a good deal of time trying to develop various models to explain fragmentary data, when only a year or two later the observations had improved so much that they showed all prior ideas, or maybe all but one, to be nonstarters. Sometimes the quality of data improves so dramatically that some fairly obvious interpretation suddenly emerges, and a consensus is quickly established.

We devote the entire next chapter to what is perhaps the most remarkable manifestation of black holes in active galactic nuclei – cosmic jets. The discovery and interpretation of jets is the fruit of a new technology – radio astronomy – that, during the last forty years, has given us a fresh picture of the Universe. The most prominent objects in the radio sky are very different from those that dominate in optical images, and the relevant radiation is emitted by quite different mechanisms. Thus, we can view the story of jets not only as the tale of a great scientific discovery, but also as a case study in how a new technology can revolutionize our understanding of a subject.

6 JETS

Narrow, fast streams of gas emerge from deep in the nuclei of radio galaxies and often project millions of light-years into space. Those galaxies are suspected to harbor black holes at their centers, and yet such jets of gas are just about the last thing you would expect a black hole to produce. Aren't black holes supposed to be (at least in popular imagination) "cosmic vacuum cleaners," sucking in everything within their reach? It seems obvious that gas being drawn toward a black hole can become hot and luminous just before it falls in, thus accounting for the luminosity of an accreting black hole. But the idea that much of the infalling gas can be "turned around" and propelled outward, at 99 percent of the speed of light or more, strains our credulity. Rather, it *would have* strained our credulity had it been proposed before the advent of radio astronomy. As has happened so many times in the past, a new technology has turned our view of the Universe on its head.

Since their discovery in radio galaxies, jets have proven to be a common phenomenon in the Universe. They often occur when gas with a lot of angular momentum swirls deep into a gravitational field, as, for example, at the center of an accretion disk. We might think of them as celestial waterspouts. They form in all kinds of systems: radio galaxies, X-ray binaries, and even ordinary stars in their infancy. So they are clearly not a manifestation of black holes alone. Yet the jets that apparently do form close to black holes carry the signature of the extreme conditions surrounding their birth – nowhere else in the Universe do we see evidence of matter propelled at such enormous speeds, sometimes to within a fraction of a percent of the speed of light.

Radio Astronomy and the Prediction of Jets

To be precise, radio astronomers rediscovered jets from active galactic nuclei. As long ago as 1917, Heber Curtis had noticed a jet of visible light emerging from the nucleus of the large elliptical galaxy M87, which lies in a cluster of galaxies in the constellation Virgo. In his 1918 publication, "Descriptions of 762 Nebulae and Clusters Photographed with the Crossley Reflector," he notes that "a curious straight ray lies in a gap in the nebulosity ... apparently connected with the nucleus by a thin line of matter. The ray is brightest at its inner end." He did not follow up the observation, nor did anyone else. Not to fault Curtis too much, we should recall that in 1918 it had not yet been established that M87 was a distant galaxy! It took another sixty years, and the development of radio astronomy, before jets were definitively established as commonplace structures in the Universe and their true natures became apparent.

Radio astronomy has its roots in the telecommunications industry. In 1931, Karl Jansky, an engineer with Bell Telephone

Karl Jansky, working at the Bell Telephone Laboratories in Holmdel, New Jersey, discovered radio waves from the Milky Way Galaxy in 1931.

Grote Reber's maps showed the Galactic Center, the supernova remnant Cassiopeia A, and the extragalactic source Cygnus A.

Laboratories in Holmdel, New Jersey, built a sensitive antenna to track down radio waves that were interfering with transatlantic telephone calls. He noticed that one of the sources of noise varied consistently with a 24-hour period; moreover, its appearance and disappearance coincided with the rising and setting of the Milky Way on the sky. Clearly, the source of these waves was not terrestrial, nor even located in the Solar System, but was tied to distant regions of our Galaxy. Jansky's announcement that he had discovered radio waves coming from thousands of light-years away startled the astronomical community, and made the front page of the *New York Times*. But curiously, it did not trigger a rush of astronomers eager to study the Universe with this radically new tool.

A few years later Grote Reber, an amateur radio enthusiast from Wheaton, Illinois, built a dish-shaped receiver in his garden specifically to look for cosmic radio signals. His maps, the first radio "sky survey," were published in *The Astrophysical Journal* in 1944. Along with the radio waves from the Milky Way, Reber found localized patches of radio emission at a few other places on the sky. One of these lay in the direction toward the constellation Cygnus, where we now know that the brightest radio galaxy, Cygnus A, is located. Radio astronomy was primarily the realm of amateurs until after World War II, when a large cohort of physicists and electrical engineers (many of whom had worked on the development of radar) re-entered civilian work at universities and government research laboratories.

The earliest radio telescopes, essentially scaled-up versions of Jansky's antenna and Reber's single dish, were severely hampered by poor resolution. As we remarked in Chapter 5, the fineness of detail (the "resolution") observable with any telescope is fundamentally limited by the ratio of the size of the aperture to the wavelength at which the observation is being made. The higher this ratio, the better the resolution. In fact, the smallest angle on the sky that can be measured by a telescope, in radians, is roughly the reciprocal of this ratio. Radio waves have long wavelengths, ranging from centimeters to many meters. By comparison, the wavelengths of visible light waves are less than one-millionth of a meter. Therefore, in order for a radio telescope to obtain the same level of resolution as an optical telescope, its "dish" would have to be between 10 000 and a million times larger! Building a single dish several kilometers across is clearly impractical, but radio astronomers have been able to exploit a powerful technique called *interferometry*, which enables a spread-out "array" of small dishes to simulate the resolution of a single large dish.

We can think of a source of radio waves as sending out ripples (consisting of electromagnetic fields) into space, much as a rock dropped into a lake sends out circular ripples of rising and falling water. A telescope

After working on radar during World War II, Martin Ryle pioneered new types of radio telescope. He invented the technique of "aperture synthesis," whereby an array of modest-sized dishes can, in combination, provide images of the radio sky as sharp as could be produced by a single dish several kilometers across. The "Cambridge surveys" played an important role in the identification of radio galaxies and quasars, as well as yielding the first significant evidence against the steady state cosmology.

effectively obtains information about the patterns of objects on the sky by measuring the directions from which the ripples are emanating. Now, if you were dozing on a rowboat in the lake and a single ripple of water came by, you would feel the boat rise and fall, but would have little clue as to the source of the ripple. But *two rowboats*, spaced sufficiently far apart on the lake, would be able to ascertain the direction of the rock by timing the rise and fall at their respective locations, and comparing notes. In particular, if the rowboats were equidistant from the source, they would rise and fall simultaneously. In a telescope, the analogous "comparing of notes" is accomplished by adding together in real time the (positive or negative) displacements of each wave, a process called *interference*. This is what happens, for example, when the radiation reflected off different portions of a mirror or radio dish are collected together at the focus. If all parts of the dish were "equidistant" from the source, all the added displacements would be either positive or negative, resulting in a large net signal.

Of course, in radio astronomy we are not dealing with a single ripple, but rather with a periodic train of ripples, spaced by a fixed wavelength. If the ripples are close together, the peaks and troughs are well defined, and timing them is easy. But if the ripples are long gentle swells, it is difficult to determine precisely when the peaks and troughs pass by. To compensate for the difficulty of perceiving ripples of longer wavelength on a lake, the rowboats would have to be spaced farther apart. Similarly, a telescope observing at a longer wavelength has to have a larger aperture in order to obtain the same angular resolution as a telescope observing at a shorter wavelength.

To address the resolution problem without building gigantic dishes, radio astronomers resort to a clever "trick." The key to achieving high resolution is to compare signals that are collected across a sufficiently long "baseline" – that is, from points that are far enough apart to discriminate the direction from which the waves are coming. It is not necessary to merge signals from every point within that baseline. In other words, to get the resolution of a 10-kilometer dish, it will suffice to mix the signals from smaller dishes that are separated by 10 kilometers. Radio astronomers can simulate the resolving power of a single 10-kilometer dish by placing a number of much smaller dishes in an array that spans 10 kilometers and using the array as an "interferometer."

But how many smaller dishes are needed, and how should they be arranged? This question must be considered carefully because directional measurements by an interferometer are much more accurate in certain directions than in others. On a lake, we can pin down the direction to the rock most accurately if the rowboats happen to be roughly equidistant from it to begin with. In a telescope, the ability of, say, a north–south pair of dishes to resolve a pair of compact sources is greatest if the sources are also lined up in a north–south direction on

As the Earth rotates, the small dishes trace out an annulus whose diameter equals the baseline separation. After combining the signals from these dishes for 12 hours, astronomers can "synthesize" the resolution of a reflecting dish with the size and shape of the annulus. By changing the separation of the two telescopes and repeating the process, one can in effect "fill in" the aperture of a single dish whose diameter is the maximum separation. To speed up the process, arrays of several dishes are used, rather than just two.

the sky. Moreover, a pair of dishes with a very wide spacing will be sensitive to fine detail, but will not be able to "see" the large-scale structure of an object. Clearly, to make a detailed picture of a radio source using an interferometer, you would need many pairs of dishes, with many different orientations and separations – and before long you'd be back to trying to build a single dish 10 kilometers across.

It was Martin Ryle, of Cambridge University, who in the late 1950s conceived a brilliant way out of this difficulty. In retrospect, it was unbelievably simple. He realized that, as the Earth rotated, the baseline of each pair of dishes would rotate with it! And as the baselines rotated, the orientation of each baseline would change with respect to sources on the sky. In effect, an array with just a few pairs of dishes having various separations could simulate a single, large dish that was much more "filled in." To build up an image of a radio source one simply had to wait and let the Earth do the work.

Throughout the late 1960s and the 1970s, a series of increasingly large interferometers was built, particularly at the University of Cambridge, at the Westerbork Observatory in the Netherlands, and at the National Radio Astronomy Observatory (NRAO) in West Virginia. These culminated, in the late 1970s, in the construction of the Very Large Array (VLA) near Socorro, New Mexico. The VLA consists of 27 linked radio telescopes, each 25 meters in diameter, distributed in a Y-shaped array spanning 40 kilometers. The VLA can uncover features that subtend an angle on the sky as small as one-tenth of a second of arc. This is the same as the angle subtended by a dime at a distance of 20 kilometers and is comparable to the resolution of the Hubble Space Telescope. We have already discussed in Chapter 5 how, in an interferometry experiment performed as early as 1953, Jennison and Das Gupta had established that the bulk of the radio emission from Cygnus A was coming not from the galaxy itself, but from two giant lobes placed symmetrically on either side of the galaxy. The new interferometers revealed this symmetric double structure to be a common feature of powerful radio galaxies. One was forced to conclude that whatever was powering the lobes had been ejected symmetrically from the galaxy.

The more sensitive radio telescopes detected additional structure. At the outer extremities of the lobes in the brightest double radio sources (the ones like Cygnus A), they often found "hot spots," compact regions emitting particularly intense radiation. Many of the brightest sources also showed tails or bridges of low-intensity emission extending backward from the hot spots toward the center of the source, where there is usually a compact focus of radio emission called a "core." Most double radio sources are associated with an elliptical galaxy (as Cygnus A is) or with a quasar; in such cases the core is invariably found to coincide with the galaxy's optical center.

The Very Large Array (VLA), near Socorro, New Mexico, uses 27 dishes, arranged in a Y-shaped pattern, to construct radio images that are as finely detailed as those produced in visible light by the *Hubble Space Telescope*.

This wealth of new detail challenged the ingenuity of theorists attempting to explain double radio sources. Initial hypotheses that the lobes were huge blobs of superhot magnetized gas, ejected in gigantic explosions, fell out of favor when it was realized that the energy would seep away as the blobs expanded, leaving too little behind to explain the observed radio emission. Some source of energy had to be replenishing the lobes so that they could continue to emit radiation. Could the hot spots themselves provide energy for the lobes, perhaps from embedded black holes flung out of the galaxy by a gravitational "slingshot"? Ingenious schemes were proposed, but an explanation for the high degree of symmetry and alignment proved elusive. It was the detection of the central cores that provided the crucial clue.

The presence of radio emission from the core suggested that there was continuing activity at the center of the galaxy. Suppose there were some kind of conduit through which this energy could be resupplied continuously to the lobes. This would solve the problem of how the energy losses in the expanding lobes were replenished and perhaps explain the symmetry as well. But what form would this energy supply take, and how could it travel with such geometric precision across so many thousands (or, in some cases, millions) of light-years of intergalactic space? Various possibilities were tried – superintense beams of electromagnetic radiation, for example – but nothing seemed to work very well. Then, in the early 1970s, researchers at Cambridge University proposed an answer that provides the basis for all current theories of radio galaxies. The lobes had to be powered by twin streams of fast-moving gas, created in the nucleus of the galaxy and propagating in diametrically opposite directions into the lobes.

To understand how the lobes and hot spots are created, one has to recognize that no galaxy exists in a vacuum. Just as the space between the stars in a galaxy is filled with an *interstellar medium* of gas and dust, so is the space between galaxies filled with an *intergalactic medium*. However tenuous this intergalactic gas may be, it will ultimately resist the free expansion of gaseous jets emanating from the galaxy. It is the presence of this intergalactic medium that leads to the observable structure of a radio source. After activity begins at the center of the galaxy, the jets propagate outward through the galaxy rather like the jet of water that emerges from the nozzle of a hose. The cosmic jets pass first through the interstellar medium and then through the intergalactic medium. As the jets progress outward, the density of the surrounding matter declines from roughly one hydrogen atom per cubic centimeter to perhaps one per million cubic centimeters. Still, the advancing jet must push the matter out of the way; hence the end of the jet moves slower than the gas flowing inside the jet. As a result, energy accumulates at the end of the jet; this is the likely interpretation of a hot spot.

The flow of gas in the jet is thought to be supersonic: that is, the speed of the gas is faster than the speed of a sound wave in the gas. The latter is much larger than the 300 meters per second attained by a sound wave in air, and may be as large as 60 percent of the speed of light. The speed of the gas is still more tremendous, in some cases reaching to within a fraction of a percent of the speed of light. As the gas approaches the hot spot, however, it decelerates suddenly. Since no sound wave can propagate upstream to warn the gas to slow down, the abrupt deceleration causes a shock wave to form across the jet. The effect of the shock is important. Before the jet reaches the shock wave most of the particles are streaming in unison, and most of the jet's energy is in the form of ordered kinetic energy. Passage through the

Schematic diagram showing the structure and mechanism of strong double radio sources. Most of the radio emission is produced in the "hot spots" and lobes, by jet material that has splattered against the interstellar or intergalactic medium.

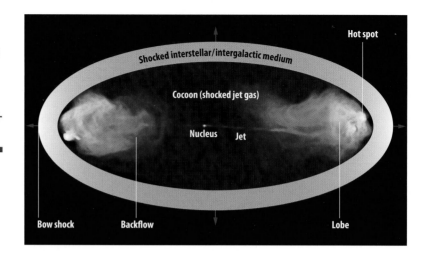

shock converts much of this ordered energy into two forms: the energy of relativistic electrons (electrons moving randomly at speeds near the speed of light) and the energy of a magnetic field. You may recognize these as the two ingredients that together yield synchrotron radiation. It is entirely natural that the most intense radio emission in a double-lobed radio source should be generated where the jet is decelerated by the action of the surrounding gas.

After splattering at the hot spots, the jet material has nowhere to go but back toward the galaxy. Thus it inflates the large lobes that can be seen in radio images. In fact the observed lobes are merely the brightest parts of a much larger, cigar-shaped bubble that envelops the entire source, called a "cocoon." Each bit of jet material spends a relatively short time (roughly 10 000 to a million years) in a hot spot before moving into the lobes or cocoon, carrying with it the relativistic electrons and the lines of magnetic field. Most of the energy of the radio source is therefore contained in the lobes and cocoon, which probably accumulate gas for 100 million years. The energy content of the lobes is prodigious, as was first realized by Geoffrey Burbidge in the 1950s. In a typical source, it must be about the amount that would be liberated if the mass of a million stars were converted entirely into energy.

Skywriting

By 1978 the theoretical arguments for the existence of jets were compelling, but astronomers were becoming uneasy. Heber Curtis's lone photograph notwithstanding, no one had yet seen any direct evidence of gaseous streams connecting the cores to the lobes of double radio

Centaurus A (also known as NGC 5128), a nearby galaxy in the southern sky (shown in visible light, (a)), exhibits huge radio lobes stretching several degrees across the sky, as well as a jet, X-ray halo, and set of "inner" lobes on much smaller scales (composite of radio, X-ray and optical observations: (b)). Both images are shown on the same scale. The energy content of the giant lobes (not shown) is colossal. Cen A was probably once a very powerful radio source; it has gradually faded as the lobes have expanded into the surrounding intergalactic medium.

(a)

(b)

This radio image of Hercules A vividly suggests the turbulent disruption of a pair of jets. The radio structure extends to a much larger scale than the galaxy itself.

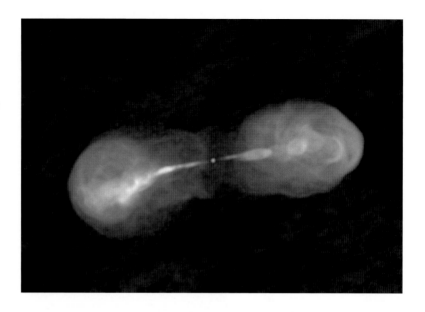

sources. But this quickly changed when the VLA came on line. In image after image, the faint traces of the energy conduits were made visible by the VLA's superior sensitivity. Soon it became obvious that jets, while ubiquitous, assumed a wide variety of appearances. In some of the brightest sources, such as Cygnus A, the jets were exceedingly faint, dead straight, and sometimes little more than a hair's width across on the image. Other jets appeared broader and brighter, but these had a tendency to "fizzle out" without producing an intense hot spot. Often the jets appeared only on one side, even when the lobes and hot spots they fed were decidedly symmetrical.

While these observations seemed to fit the prediction that jets would spurt out from the core, there was no direct evidence that the jets really represented streams of flowing gas. All we had were streaks on the radio sky that looked like they *could* be streams of gas. Therefore, acceptance of the nature of jets by most observers gained ground slowly. In retrospect, this is not surprising. Astronomers by nature tend to be conservative, and are much more trusting of quantitative evidence than of theory and attempts to draw inferences from "morphology" – or the appearances of things on the sky. (Almost the lone exception to the latter bias among eminent twentieth-century astronomers was Fritz Zwicky – co-inventor of the concept of neutron stars – whose treatise *Morphological Astronomy* is a classic of astronomical iconoclasm.) Therefore, many are very reluctant to admit the existence of a new phenomenon unless forced to by overwhelming observational evidence.

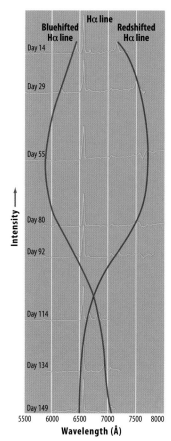

This diagram shows spectra of the X-ray binary SS 433 taken at eight different times. Curves connecting the spectra show how the observed wavelength of a particular spectral line, emitted by hydrogen, changes from one observation to the next.

The debate over the jet hypothesis reached a watershed in late 1978, when an X-ray binary in our own Galaxy, called SS 433, was found to be producing a pair of narrow jets. Unlike the jets from radio galaxies, the SS 433 jets were cool and dense enough that their composition and velocity could actually be measured, by spectroscopy. The jets were found to consist of gas with an ordinary chemical composition – mainly hydrogen – and the gas was seen to be moving at a quarter the speed of light. Here was an incontrovertible signature of the nature of jets in at least one object, albeit one that was very different from a radio galaxy.

The discovery of the SS 433 jets paraphrased, in a wry way, the discovery of the distances to quasars fifteen years earlier, when Maarten Schmidt noticed that the mysterious emission lines of quasars were ordinary emission lines of hydrogen that had been redshifted. The object's moniker pegs it as a member of a catalogue compiled by C. Bruce Stephenson and Nicholas Sanduleak, listing stars that show strong spectral emission lines. But in 1978, Bruce Margon and his collaborators at UCLA noticed that some of the lines in the SS 433 spectrum did not correspond to the wavelengths produced by known chemical elements. So far the story is reminiscent of quasars, and we know how that puzzle was resolved. But to make matters truly incomprehensible, the wavelengths of the lines in SS 433 changed with time! Eventually, Margon and his team figured out that the spectrum could be resolved into *three* sets of familiar spectral lines: one close to the correct wavelengths, one with a big redshift (that is, moving away from us at high speed), and one with a large blueshift (moving toward us at high speed). Margon reported his initial results at the Ninth "Texas Symposium" on Relativistic Astrophysics in Munich, Germany. (In keeping with the penchant of astronomers for nondescriptive terminology, the biennial Texas Symposium is seldom held in Texas. Recent venues have included Florence, Stanford, and Melbourne.) Margon's announcement was great fodder for journalists. Newspapers reported on a "mysterious dying star" that "seems to be coming and going at the same time," while Margon speculated to the press that "this is some kind of star that's in some terribly weird kind of trouble."

After about a year of observations the changes in wavelengths were found to conform to a clear pattern, which George Abell and Margon were able to fit to an extraordinarily simple model. The X-ray binary SS 433, they argued, must be producing a pair of identical, narrow jets, with speeds fixed at 26 percent of the speed of light, which precess around a fixed axis in space once every 163 days. In other words, SS 433 is a kind of cosmic lawn sprinkler, with a geometry like that shown in the figures on pages 138 and 139. But the real test of the precessing jet model would be to see the traces of the jets as they moved

The spectral lines of SS 433 come from twin jets of gas that emerge, in opposite directions, from a disk around a compact object, which is in orbit around a normal star. As the jets swing steadily around a cone every 163 days, the component of velocity directed along our line of sight varies. Because of the consequent change of the Doppler shift, the observed wavelengths of the spectral lines trace out a pattern that repeats every precession period. We see both a redshifted and a blueshifted line, with wavelengths that change with time.

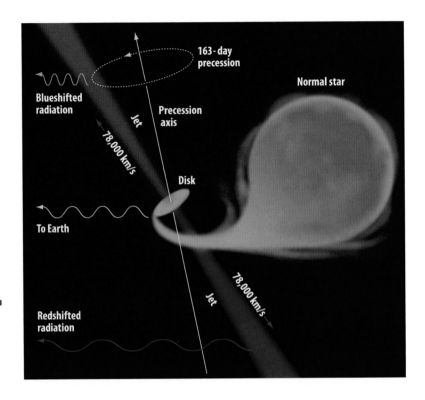

across the sky. Only 16 000 light-years away, SS 433 is close enough that the motion of the jet material should be detectable within a couple of weeks by means of a telescope with the angular resolution of the VLA. If the jets are really swinging in a wide cone once every 163 days, then we might expect to see a widening corkscrew march outward from the center, with one new turn appearing every five months. An observation like this would be a theorist's dream; indeed, many of us pinched ourselves when Robert Hjellming and Kenneth Johnston produced a "movie" of the SS 433 jets behaving exactly as predicted! Moreover, SS 433's jets seem to have produced a pair of "lobes." SS 433 lies in the center of a radio nebula, called W50. Such nebulae are usually found at the sites where stars have exploded as supernovae, and they typically have a roughly circular shape. But W50 is different. It has "ears" that are exactly aligned with the jets. Presumably, the "ears" of radio emission are places where the jets have bored most deeply into the surrounding interstellar gas. Thus, the analogy between SS 433 and giant radio galaxies seems to be a solid one.

Once we accept the idea that jets are gaseous streams, we confront a series of tricky theoretical problems. As was the case with the

Individual "blobs" of gas ejected from SS 433 travel in straight lines, but as the direction of ejection precesses around a cone, the blobs emitted at different times should trace out a widening corkscrew pattern, as illustrated here in an artist's impression. Robert Hjellming and Kenneth Johnston first observed the predicted pattern using the VLA.

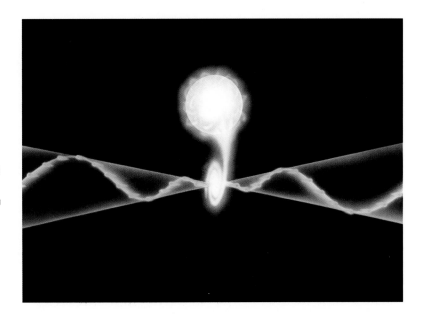

black hole paradigm for quasars, even after we understand the general concept of jets we must still answer many questions about how they behave in detail. To take one example, the observed jets seem amazingly stable against breakup. In extreme cases they seem to maintain their integrity even as they traverse millions of light-years. But everyday experience with fluid streams suggests that a jet should be highly unstable as it bores its way through the interstellar and intergalactic medium. Anyone who has handled a flexible garden hose will be familiar with this phenomenon. If the hose is laid on the ground in a curve and the

Two turns of the corkscrew pattern of SS 433's jets, mapped by Katherine Blundell and Michael Bowler using the VLA.

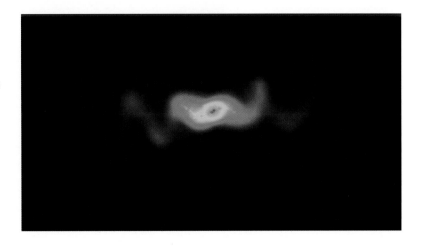

SS 433 lies at the center of this radio source, called W 50. The circular shell-like structure is typical of supernova remnants, but the two "ears" are unique. They are believed to have been produced by the pressure of the precessing jets impacting on the interstellar gas. Hjellming and Johnston's observations confirm that the axis of the precession cone lines up with the axis of the ears.

NGC 1265, the prototype of a "radio trail," is a low-powered radio galaxy in the Perseus Cluster of galaxies. It is moving through the gas in the cluster at more than a thousand kilometers per second. The high-resolution image (a) shows a normal pair of oppositely-directed jets, squirting out of the nucleus at right angles to the galaxy's direction of motion through the intracluster gas. When these jets penetrate beyond the galaxy's interstellar medium (which moves with the galaxy) they are swept into a "radio trail" ((b): low-resolution radio image in red, superimposed on optical image) that delineates NGC 1265's orbit.

(a)

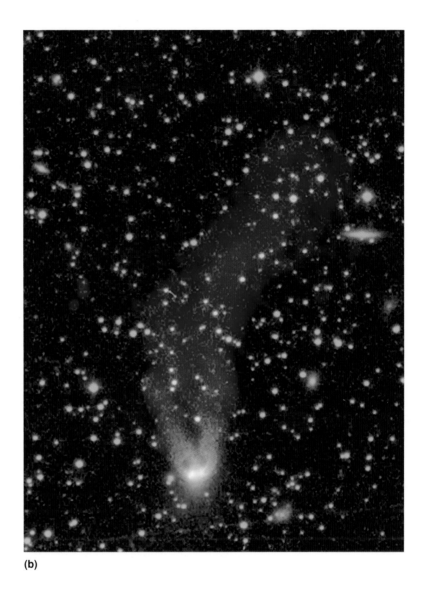

(b)

water is turned on, the hose will start to writhe. This kind of instability occurs whenever two fluids slide past each other: it is responsible
for making flags flap in the breeze and whitecaps rise on a wind-whipped
lake. We might expect it to occur along the interface between a jet and
the surrounding gas.

Clues to why cosmic jets might be much more stable than
jets in the laboratory have come from computer simulations, which
we can use to perform "numerical experiments" under conditions
completely unattainable on Earth. These have uncovered some unan-
ticipated properties of supersonic flows that may be observable in

images of radio galaxies. High-resolution three-dimensional computations incorporating electromagnetic effects have been possible for a few years, but simulations capable of handling the huge range of densities and the high speeds thought to exist in jets are only now becoming feasible.

Many jets trace out large bends on the sky. In some cases these bends may simply be manifestations of the instability described above. In a cosmic jet, however, not all bends are produced in a random manner. Consider the sources called "radio trails." At low resolution the radio emission from such a source appears not to straddle an optical galaxy as it does in a typical double source but to extend in a long curve on one side of the galaxy. A high-resolution image shows why. Two jets emerge from the center of the galaxy, but instead of terminating at hot spots the jets bend steadily to one side so that they both join up with the long, curving trail of radio emission.

What happens in radio trails is that two comparatively low-power jets are formed at the galaxy's center and are then blown sideways by the intergalactic medium as it rushes past the galaxy at speeds that may exceed a thousand kilometers per second. One is reminded of smoke that emerges from a chimney and then is caught in a high wind. The fact that radio trails are usually found in rich clusters of galaxies supports this interpretation of their morphology, for the space between galaxies in such clusters is filled with comparatively dense, hot, ionized gas. The jets feel a wind because the galaxy itself is moving at high speed through this intergalactic gas, under the gravitational influence of all the other galaxies in the cluster. It is extraordinary that the jets in radio trails can be bent through almost a right angle without losing their integrity, but numerical simulations of a jet in a crosswind show how this can happen. If jets can survive this extreme bending, it is clear that they are not always unstable. After the jets merge with the trail, their gas presumably slows down until its velocity matches the local velocity of the intergalactic medium. By observing the length of the trail and given the galaxy's velocity through the background gas, one can measure the age of the source. In a typical case the age turns out to be 300 million years.

A somewhat different explanation has been advanced for radio sources such as 3C 66B, in which the jets show several sharp bends but are otherwise basically linear. Since the bends closely mirror each other on opposite sides of the nucleus, these sources are called "mirror-symmetric." In general these sources are not found in rich clusters of galaxies, but they usually have close companion galaxies. It is likely, therefore, that we are witnessing the galactic version of a binary star system, with the radio galaxy and its companion in orbit around each another. If the velocity of the gas in the jets is not much

This composite radio/optical image shows the radio trail NGC 4869, plunging through the outskirts of the Coma cluster of galaxies. Closer to the cluster's center, NGC 4874 exhibits a "wide-angle tail" of radio emission, possibly due to buoyancy or sloshing motions of the intergalactic gas.

greater than the orbital velocity, the orbital motion of the galaxy will be "written" on the sky in the form of bends in the jets. In a sense the bends are an illusion: the gas in the jets does not flow through them. Imagine a photograph of someone watering a garden by swinging a hose. The water hanging in the air has a zigzag pattern, but a given drop of water in the pattern is moving in a ballistic arc. Analogously, a given particle of gas in a jet follows a straight path. The shape of the bends in a cosmic jet can be used to infer the trajectories of the orbiting galaxies. Note that orbital motion will automatically produce a mirror-symmetric pattern on the sky, exactly as we see in objects like 3C 66B: if one jet bends to the right, the other jet must also bend to the right at about the same distance from the center of the galaxy.

A third class of "bent" radio sources exhibits inversion symmetry. In this case a bend to the right in one jet corresponds to a bend to the left in the other. An X-shaped pattern like this could result from a sudden change in the direction of the jets, caused by a flip in the direction

The pattern of radio emission in 3C 66B displays mirror symmetry, which might result from the orbital motion of the galaxy about a companion.

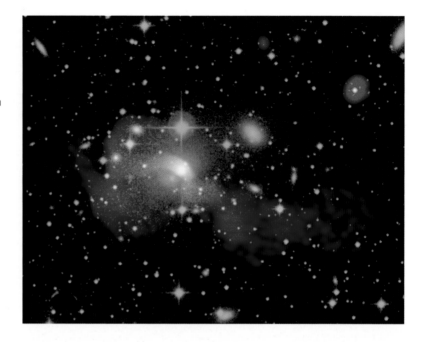

of the black hole's spin, following a merger, or by a sudden change in the angular momentum of the gas feeding the disk. This pattern could also be produced if the jets were precessing like a top, just like SS 433 but on a much longer timescale – millions of years instead of 163 days.

Approaching the Speed Limit

If the properties of jets on large scales seem surprising enough, their perceived nature becomes even more exotic as we explore closer to the nucleus. But on this exploration even the powerful Very Large Array fails us, since its angular resolution is limited to one-tenth of an arc-second, or about 300 light-years at the distance of Cygnus A. If there is a black hole at the center of Cygnus A, its event horizon must be more than a million times smaller! Fortunately, radio astronomers have devised instruments that can resolve details up to *a thousand times finer* than the VLA.

As we explained earlier, the resolution of maps made by radio interferometers is limited by the separation of the individual radio dishes: the greater the separation, the greater the resolution. To perform interferometry one must be able to combine the signals from the indi-vidual receivers. Across separations as great as 40 kilometers, like those between the VLA dishes, underground wave guides can link the dishes in real time. Direct linkage with the required precision is unwieldy for

The inversion symmetry seen, for example, in this radio image of the galaxy NGC 326, could have arisen if the spin axis of the black hole suddenly changed direction following a merger with another black hole. The high-resolution *Hubble* image (inset) shows the current orientation of the jets – prior to the merger they would have been aligned with the outer lobes.

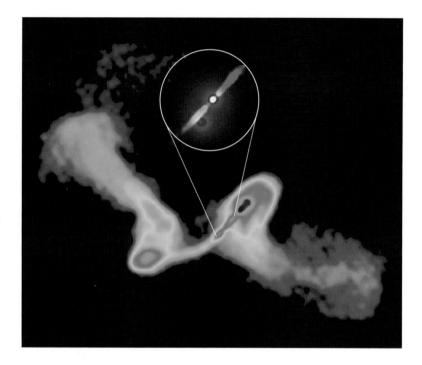

much larger arrays, although Great Britain's "MERLIN" array of radio telescopes (headquartered at Jodrell Bank, near Manchester) achieves resolutions about 10 times higher than the VLA by using optical fibers to link antennae spread over several hundred kilometers.

In the technique called *very-long-baseline interferometry* (or VLBI), which was developed simultaneously with the construction of the modern generation of linked radio interferometers, the receivers can be on different continents. The resolution that can then be achieved is correspondingly greater. Features as small as a ten-thousandth of a second of arc can be distinguished – at this resolution, you could read a newspaper headline from across the Atlantic. The angular scale that can be resolved corresponds to one light-month at the 300 million light-year distance of the radio galaxy NGC 6251 and 3 light-years at the distance of the farthest quasars. This great resolution is achieved at a cost, however. The individual telescopes cannot be linked in real time; hence the radio signals from the individual receivers must be recorded separately (but simultaneously) and compared later, using a special-purpose computer called a *correlator*. This procedure entails a loss of information, so that one has to work harder (and observe longer) to obtain images of comparable quality to those made with shorter-baseline interferometry. In the early days of VLBI, the opportunities to make high-sensitivity observations were few, since VLBI measurements were made on an ad hoc basis

The Very Long Baseline Array (VLBA) consists of 10 radio dishes extending 8000 kilometers across the United States from Hawaii to the Virgin Islands. Operated as an interferometer, the array serves as a "telescope" with the same resolution (though not the same sensitivity!) as a single dish the size of the Earth. The Y denotes the location of the VLA, which provides additional data, with lower angular resolution but greater sensitivity, to supplement the VLBA.

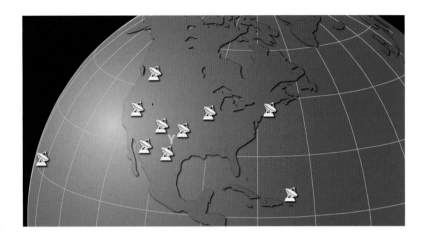

by combining data from radio telescopes (built for other purposes) around the world. But this changed when the Very Long Baseline Array (VLBA), a system of 10 dishes spanning the United States from Hawaii to Puerto Rico, came online in 1993. The VLBA allows observations to be made full-time and yields high-quality images on a routine basis.

Very-long-baseline interferometry has been notably successful in probing the cores of radio galaxies. It has shown that they usually have features that remain unresolved. These features must therefore be smaller in angular size than a few thousandths of a second of arc (corresponding typically to a few light-years). In contrast, the hot spots in the extended lobes generally have no small-scale structures. VLBI has also been successful in studying a separate class of objects called compact radio sources, usually found at locations where optical telescopes reveal distant quasars. In each compact radio source, the radio flux comes mostly from a small core, not from extended lobes. Some additional radio emission is also generally detected at a low level from an irregularly shaped region surrounding the core. The compact radio sources are often found to vary in intensity on timescales ranging from months to years, but some are found to vary within a few hours! We shall revisit these objects later.

By comparing high-resolution VLBI observations with broader images made using linked interferometers, we can construct an "exploded view" of a radio galaxy showing its structure on different angular scales. With these composite images we can "follow" jets all the way from the largest scales of the radio lobes into the innermost few light years of the galactic nucleus. The figure of M87 as seen on different scales provides a striking example. The inner region around the nucleus typically exhibits a single jet emerging from an unresolved point of intense radio emission. This has been dubbed a "core-jet" structure. In many instances the single jet points straight in the direction of one of the source's radio lobes,

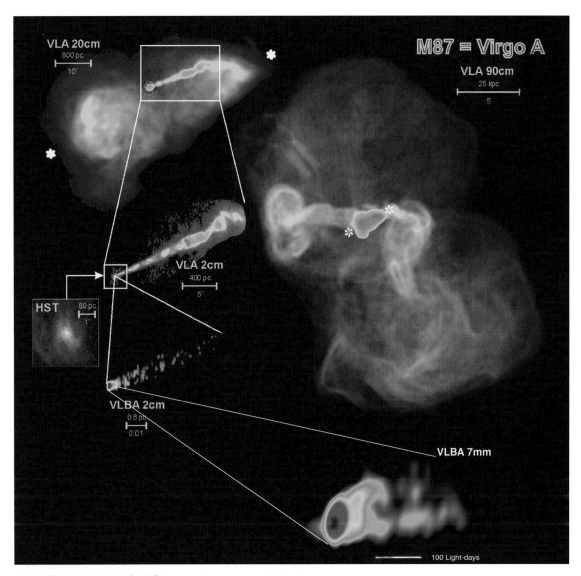

By combining images made with different radio telescopes, Frazer Owen and his colleagues have traced the structure of M87 on all scales from a quarter-million light-years down to a light-month, which corresponds to 100 Schwarzschild radii for the estimated mass of M87's black hole.

although the jet sometimes curves away from this direction on the smallest scales. Although only a single jet is visible, it is possible that a second jet points in the opposite direction but is too faint to be seen. The fact that a jet appears on these smallest scales provides the strongest evidence that the lobes are indeed supplied with energy through the jets. The activity in the core must have been continuing for tens of millions of years for the lobes to have accumulated their energy from these jets. The continuous alignment of the jet all the way from the nucleus to the lobe indicates that its direction has remained constant over that time.

When we turn to the broader picture of the radio emission obtained by linked interferometers, we often still find only a single jet, on the same side as the jet seen in the small-scale images of the core. This one-sidedness often persists out to very large distances from the nucleus, more than a million light-years in some extreme cases. Does the detection of only a single jet mean that all the energy is coming out on one side? Perhaps, but this can't be true for all time. Since the lobes are roughly symmetric, the energy output averaged over time must be essentially the same on both sides. To avoid this objection one might suppose that the jet flip-flops from side to side on a timescale that is relatively short compared to the roughly 100 million years it took to create the lobes. But the continuity of one-sidedness from small to large scales argues against the flip-flop theory, since one would expect to see sources that have flipped between the time the large-scale structure was created and the time we are viewing the jet on small scales.

Of all the debates over jet physics, few have proven so lively as the argument over what causes the "sidedness" of jets. Above we tried to give a flavor of the kinds of arguments that were made early on in the debate. But now we must bring in another, even more surprising observational trend. It suggests a simpler, albeit more exotic, explanation for one-sidedness.

It turns out that the small-scale structure of jets that can be mapped with VLBI often changes from year to year, in a systematic way. From time to time bright features in these jets – often called "knots" or "blobs," for want of a better term – appear to form close to the unresolved core and then to march outward with time. Taking into account the distance to the radio source, as measured using the redshift of the galaxy, we can estimate the apparent speed at which the "blobs" move outward on the sky. Surprisingly, the features move at a speed that seems to be in excess of the speed of light; this phenomenon is therefore known as *superluminal expansion*.

Astronomers do not contend that these "blobs" are really moving faster than light. Instead, it is thought that we are witnessing an illusion. The bright patch appears to be moving faster than light because the radiation from the patch and the radiation from the core of the galaxy, which serves us as a benchmark, travel along paths of different lengths and therefore take different amounts of time to reach us. The simplest geometric arrangement requires that the patch be moving toward us at a small angle to the line of sight between the radio source and the Earth, with a speed slightly less than the speed of light. Radiation from the core always takes the same amount of time to reach us, whereas the time it takes for radiation to reach us from the patch decreases with time. This makes the patch *seem* to separate from the core at an apparent rate that can be far greater than the true rate and indeed greater than the speed of light. The important point is that superluminal expansion does not imply that blobs in radio

High-resolution radio images of quasars typically show a bright core and a one-sided jet with a blobby structure. This series of images of 3C 279, made between 1991 and 1998 by Ann Wehrle and her collaborators, reveals noticeable changes in the structure of the jet. The (blue-green) "blobs" are marching outward, while new features occasionally appear near the core and join the march. The blob near the end of the jet has "moved" nearly 2 milliarcseconds, corresponding to more than 25 light-years at the distance of 3C 279. This kind of "superluminal expansion" is observed in many compact radio sources; its explanation requires the existence of outflows pointed nearly at us with speeds only fractionally less than that of light.

jets are traveling at speeds faster than the speed of light, but does require that they travel at speeds *very close* to the speed of light. For example, the superluminal expansion at 10 times the speed of light observed in 3C 273 requires blob speeds exceeding 99.5 percent of the speed of light.

Once we accept the idea that the speeds of gas flow in jets are *relativistic* (that is, very close to the speed of light), the puzzle of one-sidedness is immediately solved. To explain why, we must introduce

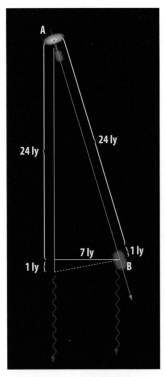

The geometry of superluminal expansion. If the blob in this example moved exactly at the speed of light, it would appear to have a transverse speed of 7 times the speed of light. It would actually take the blob 25 years to move from point A to point B, but light emitted when the blob starts out at point A would reach us only one year before the light signaling its arrival at point B. The blob would therefore appear to make the trip in only one year! Nothing can, of course, move at exactly the speed of light. However, this example shows that an object moving at 96 percent (24/25) of the speed of light would appear to move 7 light-years transversely in just over 2 years, implying a transverse speed nearly 3.5 times that of light.

the phenomenon of *relativistic aberration*. Suppose a hunter wants to shoot a duck when it is directly above. If the shotgun is aimed vertically upward, it must be fired before the duck passes overhead. The shotgun pellets will travel vertically upward to the duck, which will have moved forward to meet them (the hunter hopes). Now consider all of this from the vantage point of the duck. It sees the hunter moving toward it, and the pellets, instead of moving vertically, have a horizontal component in their motion. To put it another way, the duck sees the pellets travel along a slightly aberrant trajectory that is inclined toward the direction in which the hunter is moving with respect to the duck.

The same thing happens with photons, the "particles" of electromagnetic radiation. A blob of hot gas that radiates photons equally in all directions will appear to be shining preferentially along its direction of motion. When the blob is moving relativistically (that is, almost as fast as the photons), the effect is so pronounced that it is sometimes called the "headlight" or "beaming" effect. Take the case of a blob of gas in a radio jet whose relativistic motion gives the illusion of superluminal expansion at an apparent speed of five times the speed of light. Here roughly half of the photons that the gas emits will be radiated within a cone facing into the direction of its motion with an opening angle of only 5 to 10 degrees. Furthermore, each of the photons within the cone will have been made more energetic by a Doppler-shift decrease in its wavelength (since the energy of a photon is inversely proportional to its wavelength). The net result is striking. If an observer is in the cone, the source will look brighter than it would if it were stationary, typically by a factor between 100 and 1000. Conversely, if the observer is well outside the cone, the source will be effectively invisible. The fact that the jet is seen only on one side of the galaxy most likely indicates that its flow is relativistic; unless the axis were precisely in the plane of the sky, the Doppler beaming effect would brighten the jet on the approaching side.

These ideas have motivated astrophysicists to develop so-called unified models for radio galaxies. We alluded to these models in Chapter 5. The basic idea is that radiation sources of a standard kind, if they emit in a directional ("anisotropic") way, can have very different appearances depending on how they are oriented relative to our line of sight. Suppose most radio sources consist of a pair of relativistic jets emerging in opposite directions from the heart of a galactic nucleus. If the jets are more or less aligned with the direction from the source to the Earth, we detect the jet that is pointed toward us; the other jet is invisible. We therefore see a bright, one-sided jet, foreshortened by the small viewing angle. The jets end in a pair of extended radio lobes, which radiate photons roughly equally in all directions because the relativistic flow is drastically slowed when the jets pass into the hot spots. However, because we are viewing the source "end-on," the lobes appear to us as a relatively faint halo of extended

Interstellar clouds

Supernova

All photons scattered off this
surface would be delayed
by the same amount of time

In the years after Supernova 1987A exploded, two rings of light were visible around it; their radii appeared to be expanding at several times the speed of light. This observation, an example of a "light echo," does not mean that Einstein's theory of relativity is being violated. The apparent super-light speed resulted from the geometry responsible for producing the rings. The flash of light from the explosion was being scattered off dust particles concentrated in two clouds lying in front of the supernova in the Large Magellanic Cloud. A similar geometric arrangement leads to apparent "superluminal" expansion of compact radio sources.

emission that is seen to surround the strongly beamed compact source. This is the most likely explanation of the compact sources described earlier: they are probably those big double sources that are seen nearly end-on.

In the radio galaxies that show extended lobes, the jets are presumably traveling at a large angle with respect to our line of sight. For the most powerful of these sources, the ones such as Cygnus A, the jets are highly relativistic and extremely faint. For sources somewhat less powerful, the jets are mildly relativistic. Nevertheless, the aberration of their radio emission is sufficiently pronounced that only the jet on our side of the galaxy's center is bright enough to be detected. For weak sources such as 3C 66B the jets are significantly slower than the speed of light on the large scales observed by linked interferometers such as the VLA, so that both jets are equally visible. We note, however, that even in these symmetric-jet sources the jets appear one-sided on small VLBI scales, so that we are apparently looking at jets that start out at near-light speeds and are slowed down by their interactions with the interstellar gas.

The ultimate test of a good scientific theory is its ability to account for observational data that were not used to arrive at the model. "Unified models" for radio galaxies pass this test with flying colors. One generic prediction of unified models is that the jet we see in a one-sided source should be the one on the near side of the galaxy. Since galaxies contain interstellar gas this means that the radio lobe on the "jetted" side of the galaxy should have much less gas in front of it than the lobe on the "unjetted" side. Several years after unified models were proposed, Robert Laing and Simon Garrington, of Cambridge University, found exactly this effect in nearly every source they measured.

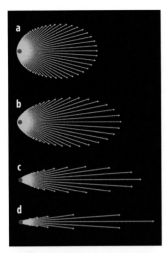

When an object moves at nearly the speed of light, relativistic aberration concentrates the radiation it emits into a "forward cone" around its direction of motion. In (a), the emitter (a cloud that intrinsically radiates equally in all directions) is moving toward the right at half the speed of light. In (b) it moves at 75 percent the speed of light, in (c) at 94 percent, and in (d) at 98 percent the speed of light. The emitter is rendered almost invisible except from the forward direction. The shape of each pattern shows how the intensity varies with the angle of emission. Seen from directly in front, the emitter appears brighter than if it were stationary by factors of 7, 30, 440, and 3100, for the four cases (a) to (d) respectively. (These enormous enhancements are, of course, not shown to scale in the figure.) Aberration leads to a strong selection effect in favor of detecting relativistic jets pointing almost toward us – precisely the ones that manifest apparent superluminal motions.

Another correlation that neatly falls into place concerns the alignments of small-scale jets observed with VLBI. Anthony Readhead, Marshall Cohen, and their colleagues at the California Institute of Technology discovered that in certain radio galaxies the jets align neatly with the lobes. These are precisely the galaxies dominated by extended radio emission – that is, whose emission comes mostly from double lobes. In these sources, any misalignment is by no more than a few degrees. In contrast, the VLBI jets associated with compact radio sources – in which the core dominates the emission – are noticeably curved, so that they are seriously misaligned with the faint extended structure of the lobes, typically by angles of 30 degrees or more. This correlation is easily explained by unified models. One does not expect the jets to be precisely uniform and straight – doubtless they exhibit some natural bending and inhomogeneity. In compact radio sources, which we are viewing end-on according to unified models, the bending is exaggerated because our line of sight is almost straight along the jet. Moreover, we would expect superluminal motion to be most pronounced in the compact sources, as indeed the observations seem to show.

A third correlation concerns the variability of radiation emitted by AGN. When we were discussing the black hole paradigm for AGN in Chapter 5, we used variability to place upper limits on the size of the energy-emitting region. But some kinds of variability are incompatible with this simple interpretation. Striking results have been obtained at gamma-ray wavelengths, by the *Compton Gamma Ray Observatory* (a NASA satellite that orbited the Earth between 1991 and 2000) and, more recently, by ground-based telescope arrays observing the impacts of very-high-energy gamma-rays on the Earth's atmosphere. Some compact radio sources appear to be emitting a large fraction of their total luminosities in the gamma-ray portion of the spectrum. Moreover, these gamma rays sometimes fluctuate within a matter of days or less (the current record is a few minutes, for observations made with a groundbased array). If the timescales on which the gamma rays vary were assumed to represent the light travel times across the emitting regions, the sources would be so tiny that the radiation could not escape. But unified models offer a way out. Just as relativistic motion nearly along the line of sight can lead to the illusion of superluminal expansion, it can also lead to apparent variability that is much faster than the true variability. Therefore, the emitting region can be much larger than suggested by variability, and the escape problem goes away.

The "Doppler beaming" interpretation of extragalactic radio sources accounts for so many diverse observations that it is unlikely to be seriously wrong. Thus, radio astronomy leads us to some profound generalizations about the nature of activity in the nuclei of galaxies. Those AGN that produce intense radio emission – radio galaxies, BL Lac objects, and a subset of quasars – invariably do so through the production

The two oppositely directed jets emerging from the nucleus of the radio galaxy 3C 31 are not exactly in the plane of the sky. We observe more intense radiation from the jet directed toward us because of the Doppler effect. The apparent asymmetry implies jet speeds more than 90 percent that of light.

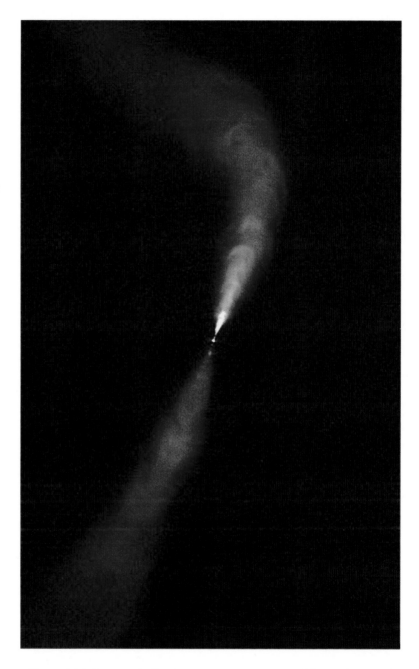

of jets, which by every indication appear to start out with speeds exceedingly close to the speed of light. Where do these jets come from? Even VLBI cannot resolve the sources of jets. But what we can say is that jets are already well formed on scales that are smaller than those accessible to VLBI. In the case of the nearby radio galaxy M87, which is "only" about

50 million light-years away, this means scales smaller than 0.03 light-year, or less than 100 times larger than the likely size of the black hole.

Making a Relativistic Jet

Hidden from direct observation by its small scale, the process of jet formation in AGN is very much the playground of theorists. We think we know something about the environment in which a jet forms, and we can see the end product as the jet emerges from the nucleus. But everything else about the process must be deduced from first principles. What elements must go into a theory of jets? First, there must be a source of energy that can propel gas to high speeds. Second, there must be a mechanism that splits this energy flow into two equal branches (since jets come in pairs) and gives these flows a specific direction. Third, there must be some process that focuses the twin flows into the narrow streams we see. We can be reasonably confident that the jets are initially propelled along the rotation axis of the "central engine." But the natures of the first and third elements are much more uncertain.

Creating a flow that approaches the speed of light is especially challenging. The easiest way to accelerate gas to high speed is to push it with some form of pressure. If you heat gas to a high temperature in a confined space and then let it go, the pressure will accelerate the gas as it is released, like air being released from a balloon. If all the thermal energy goes into acceleration, this process is called *adiabatic expansion*. The final velocity reached by the accelerated gas will then be of the same order as the typical random particle velocity – roughly equivalent to the sound speed – that the gas had when it was hot. But heating and releasing gas turns out *not* to be a good way to attain speeds close to the speed of light. To reach relativistic speeds by this kind of mechanism, one would have to heat the gas to such a high temperature that the thermal energy per particle would greatly exceed the "rest-mass energy" ($E = mc^2$) per particle. Gases at such high temperatures tend to be unstable. There is so much energy present that it is very easy for some of the energy to be turned into new particles! This energy is then not available to help with the expansion. Such hot gases also tend to cool very rapidly, which means that they radiate away their energy (or even emit other escaping particles such as neutrinos) before it has a chance to contribute much to the acceleration.

Another way to accelerate gas in AGN is to push on it with radiation. But the relativistic aberration effect, discussed earlier, limits the ability of radiation pressure to accelerate gas to the highest speeds that have been observed. Suppose you were riding on a blob of gas that is being accelerated by a beam of radiation coming from behind. Realistically, the light rays pushing on the gas cannot be perfectly aligned; they will have at least a small spread in direction. As the gas approaches the speed of light, the aberration effect causes the light rays on the edge of the beam to appear as if they are coming at you from the front! These rays *oppose* further acceleration and will generally limit the maximum speed. There is a way out of this paradox, however, if the accelerating gas is so dense that it *traps* the radiation, progressively focusing it into a cone about as narrow as the Doppler beaming angle associated with the motion of the gas. The conditions required for trapping are unlikely to apply in AGN but may exist in gamma-ray bursts, as we shall see in Chapter 7.

In addition, pushing on a gas with radiation cannot yield a final jet flow with more energy than was contained in the radiation to begin with. Unless most of the radiation energy can be "used up" during the acceleration process or the radiation source is hidden from view, any jet source powered by radiation pressure should also be very luminous. Yet strong radio galaxies, such as Cygnus A and M87, are not luminous enough. To energize the extended radio lobes, the jets would have to have more power than is available in the radiative luminosity of the nucleus.

Is there a mechanism that could accelerate a powerful outflow of gas to relativistic speeds, without producing an intense source of radiation? For the past three decades, attention has focused on the role of magnetic fields. In almost all ionized gases encountered in astrophysics, there are enough electrically charged particles present to "short out" any significant electric field. But there is no way to neutralize magnetic fields. One can then describe the electromagnetic properties of an ionized gas in a very simple way, using a viewpoint first introduced by the Swedish physicist Hannes Alfvén. We can think of the magnetic field lines as being part of the fluid, in the sense that when the fluid moves, the magnetic lines of force must move with it – and vice versa. The embedded magnetic field also gives a fluid added resiliency. If the field lines are crammed closer together, they will try to spring back; if they are gently curved, they will try to straighten out; but once curved into a closed loop they will try to contract like a stretched rubber band.

Magnetic field lines passing through an accretion disk will be forced to rotate with the disk. The particles of ionized gas above and below the disk are then forced to rotate along with the field lines, and consequently feel a centrifugal force that flings them outward. Beyond a certain distance, however, the magnetic field is not strong enough to force the particles to corotate. The gas particles lag behind, bending the magnetic field lines backward. As the field lines become more and more twisted about the rotation axis, they begin to exert a strong pinching force that may be responsible for the initial focusing of the jet. This tendency of a helical magnetic field to contract toward its axis could explain why jets are so narrowly focused, or collimated, even on the smallest observable scales. Under certain circumstances the coiled

In the magnetocentrifugal mechanism for accelerating and collimating jets, magnetic lines of force pierce the accretion disk and rotate with it (left). If the magnetic field lines make an angle of less than 60 degrees with the disk plane, gas rotating along with the field is flung outward by centrifugal force. When the gas reaches a high enough speed, its inertia twists the magnetic field into a helix (right), which exerts a pinching force that focuses the gas into a collimated jet.

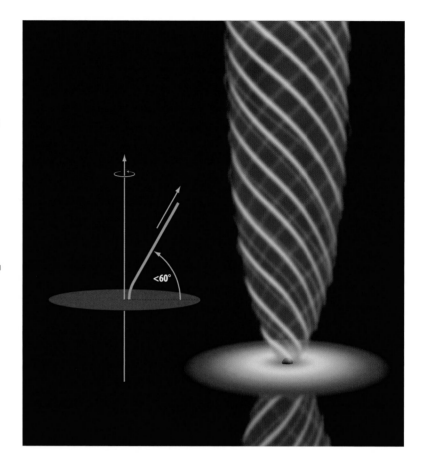

magnetic field can also act like a compressed spring, helping to accelerate the gas still further.

The "magnetocentrifugal" process is especially attractive because it seemingly explains both the acceleration and the collimation of jets using a single mechanism. It is also appealing because it is potentially applicable to a wide variety of systems. We now realize that jets are ubiquitous in the Universe, and often do not involve black holes or compact objects of any sort. Even the swirling gaseous disks within which stars like the Sun are forming produce strikingly narrow jets, which have been traced over several light-years in length. As is the case for the relativistic jets in AGN, neither hot gas nor radiation pressure seems adequate to propel these "protostellar jets" to the observed velocities. Once again, speculation centers on magnetic forces as both the driving and collimating agents.

Protostellar jets are very different from their AGN counterparts. Their velocities are "only" about 100 kilometers per second (1/3000 that of an AGN jet), and they may even contain molecules. Yet they can be understood according to the same general principles as the

Despite its eerie resemblance to a giant double radio galaxy, this is a visible-light image of the region around a young star, made using the *Hubble Space Telescope*. The protostar itself is hidden behind a dusty cloud, but its jets penetrate the cloud and create a pair of lobes spanning more than a light-year. Theorists believe that magnetic propulsion and focusing mechanisms, similar to those operating in AGN, can also explain protostellar jets.

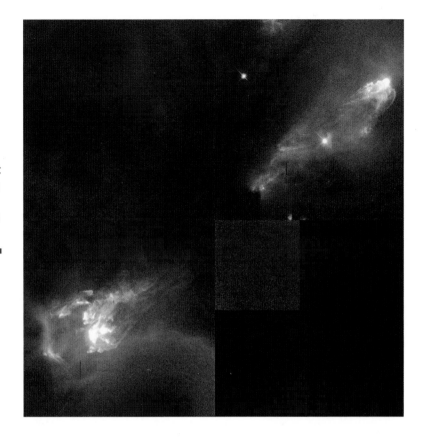

jets from spinning black holes: gaseous disk rotates magnetic field lines, material is lifted off disk and flung outward by centrifugal force, twisted field pinches incipient jet, and so on.

How far should one take this analogy? The source of energy for a protostellar jet is presumably the accretion disk itself; not surprisingly, these jets achieve speeds similar to the escape speed from the inner parts of the disk. Jets in AGN achieve speeds far higher than the maximum orbital speed of the disk: they are accelerated to within a fraction of a percent of the speed of light. Could such a flow be produced using only the gravitational energy released by infalling matter, as happens when the material in an accretion disk swirls into a compact stellar remnant? An accretion disk may dispose of some of its energy by creating a wind; it is an open question whether such an outflow can reach the "ultrarelativistic" speeds of AGN jets. But there is another source of power that is available to be tapped: the rotational energy of a spinning black hole.

When this possibility was first discovered by Roger Penrose of Oxford University, it came as a shock, since people had thought that *nothing* could be extracted from a black hole. But this is not quite true.

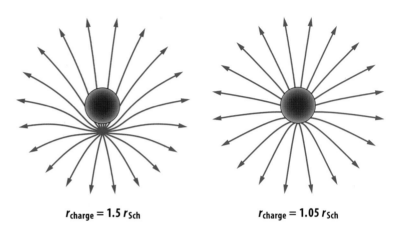

$r_{charge} = 2.5\, r_{Sch}$

Electric field lines

Nonrotating black hole

Point charge

$r_{charge} = 1.5\, r_{Sch}$

$r_{charge} = 1.05\, r_{Sch}$

If an electric charge is lowered toward a (Schwarzschild) black hole, the lines of force become distorted. As the charge gets close to the Schwarzschild radius (or "horizon") the distortion becomes so great that the lines of force appear to emanate from the center of the hole. It is as though the horizon were a conducting sphere over which the charge spreads itself uniformly. If the point charge is falling into the hole, the apparent spreading occurs on the free-fall time. This makes it appear as though the horizon has an electrical resistance of about 100 ohms.

The real law governing the growth of black holes requires only that the *surface area* of a black hole's horizon can never decrease. Since a spinning hole has a smaller surface area than a nonrotating hole of the same mass, there is no reason in principle why the rotational energy could not be extracted. But how does one do it in practice?

Astrophysicists have proposed plausible mechanisms for extracting the spin energy of a black hole; these mechanisms depend on exploiting the remarkably close analogy between a black hole and an ordinary electrical conductor. This analogy can be illustrated for a nonrotating, or Schwarzschild, black hole (discussed in Chapter 1) as follows. Imagine observing the electric field around a point charge that is gradually moved toward the hole. When the charge is far away from the hole, the electrical lines of force stick out radially from the charge. But as the charge is moved closer to the horizon, the lines of force get progressively more distorted: they "wrap around" the hole so that eventually they appear to emanate radially from the center of the hole instead of from the charge. It is as though the charge had spread itself over the hole's "surface," just as would happen if the "surface" were an electrical conductor.

A spinning, or Kerr, black hole behaves like a spinning conductor. Suppose that the gas approaching a spinning black hole contains a magnetic field. Like any spinning conductor in a magnetic field, the "surface" of a Kerr black hole will develop a voltage difference between the equator and poles. This means that a black hole spinning in a magnetic field behaves like a battery, analogous to the battery in a car or flashlight. In electrical engineering parlance, this kind of a battery – a conductor spinning in a magnetic field – is called a *unipolar inductor*. The only difference between a black hole battery and a car battery is that the voltage drop across the poles of a black hole battery in an AGN would be around 10^{15} volts, instead of the more familiar 12 volts!

Ionized gas

Electric current

Rotating black hole

Magnetic field lines to equator

Magnetic field lines to pole

A spinning black hole behaves like a spinning conductor. If the hole is surrounded by conducting gas carrying a magnetic field, then a voltage difference is set up between the poles and the equator, and the hole can act like a battery. The electric current driven by the voltage drop can extract energy from the spin of the hole, which can be transformed into outflowing jets.

To extract energy from a battery, you have to hook it up to a circuit. In the case of a black hole battery, one "wire" should be attached to the pole and the other to the equator. In an AGN, the role of the "wire" is played by the ionized gas that surrounds the black hole, since this gas is an excellent conductor. The amount of energy extracted from the battery then depends on the relationship between the electrical resistance (or "impedance") in the circuit and the internal resistance inside the battery. The power output is maximized if the two resistances are equal, a condition called an *impedance match*. Now it turns out that a black hole is not a perfect conductor. No matter what its mass, it behaves as if it has a resistance of about 100 ohms, roughly the same resistance as the light bulb in a flashlight. But what is the external impedance? If the effective circuit closes at some significant distance from the hole, as we expect, then the impedances will automatically be roughly matched. Energy, drawn from the hole, accelerates charged particles and ultimately propels the jet. As the energy is extracted, there is a drag on the spin of the hole, slowing it down. In this sense, we might think of the spinning hole as a kind of flywheel, storing rotational energy for later use.

The huge voltage drop across a "black hole battery" may actually be responsible for *creating* the matter that flows along a jet. In the presence of such a large voltage, any electron is accelerated to such a high energy that it creates a "cascade" of particles of matter and antimatter – mostly electrons and positrons – from the vacuum of space. One could even say that the black hole "short circuits," and compare the creation of particles to the energy released in a bolt of lightning, when the voltage between a thundercloud and the ground "shorts out."

Basic features of the black hole spin extraction mechanism are beginning to be confirmed by computer simulations, albeit under many simplifying assumptions. According to these calculations, an extremely relativistic flow is produced in the evacuated funnel surrounding the rotation axis, with a slower wind boiling off the surface of the accretion disk. The energy output in the jets increases steeply with the spin of the black hole. How fast do we expect a supermassive black hole to be spinning? It may, of course, have "original spin," because it formed by the collapse of a rotating gas cloud or supermassive star. Moreover, any black hole in an active galactic nucleus can be "spun up" after a sufficiently long time by accreting material with angular momentum. A third way to obtain a rapidly spinning black hole, in shorter order, is from the merger of two holes with comparable masses. In that case, a sizable fraction of the orbital angular momentum ends up in the spin of the merged hole. It is more difficult to predict what the magnetic field will be like in the gas surrounding the hole. In accretion disks,

Roger Blandford has contributed several key ideas on the nature of active galactic nuclei. In particular, he studied the formation of jets, and with Roman Znajek showed how the spin energy of a black hole embedded in a magnetic field could be converted into an outflowing jet.

In this snapshot from a computer simulation by Shigenobu Hirose and collaborators, strands of "spaghetti" trace the structure of the magnetic field in an accretion flow onto a spinning black hole and in the associated jet. The background color shows the gas density varying by a factor of 10 000, from high values (purple) to low (orange).

powerful instabilities tend to amplify the field to a sufficient *strength* that efficient energy extraction could occur. But it is not known whether the *structure* of the field will favor energy extraction. Some simulations indicate that the swirling of the gas in its final plunge toward the black hole can help to set up favorable conditions by organizing and intensifying the field.

There is certainly no observational evidence that the conditions for jet formation are satisfied in all cases. Indeed, only a small fraction of AGN produce powerful jets. This could mean that rapid black hole spin is rare – a prediction of some black hole formation scenarios – or that energy extraction is sensitive to details of the magnetic field structure. Curiously, in cases where the galaxy hosting the AGN can be identified by type, the "radio-loud," jet-producing nuclei seem to be in ellipticals, whereas the radio-quiet nuclei are in spirals. We think of the spiral vs. elliptical classification as referring to the gross properties of galaxies, whereas the question of whether or not to form a jet is decided on scales not much larger than the horizon of the black hole. How the innermost nucleus "knows" in which kind of galaxy it resides is one of the principal puzzles in AGN research.

Nor is it proven that the extraction of black hole spin energy is the only mechanism that can accelerate jets to relativistic speeds. At least one X-ray binary believed to contain a neutron star, Circinus X-1, produces a one-sided jet that exhibits superluminal motion. If the neutron star identification is correct, it would mean that the electromagnetic processes peculiar to black hole event horizons are not required to produce ultrarelativistic flows. Perhaps unbalanced magnetic forces, operating in a tenuous corona above the accretion disk, could channel enough energy into a small amount of matter to propel it to speeds close to the speed of light. Or, some process involving the neutron star's surface, spin, or intrinsic magnetic field might mimic the effects of black hole spin.

Judging from the substantial radio lobes inflated by Circinus X-1 in the interstellar medium, such a mechanism would have to be very efficient, pumping roughly as much energy into jets, over time, as the accretion disk puts into radiation. Circinus X-1 is a member of a class of jet-producing X-ray binaries, called *microquasars*, that includes the archetype of black hole stellar remnants, Cygnus X-1 – which also has inflated a large lobe. The discovery of microquasars answered a longstanding question about the *scale invariance* of the mechanisms that produce jets. Microquasar jets resemble miniature versions of the jets produced by extragalactic double radio sources. They may result from similar processes, but are energized by a black hole of around 10 solar masses, rather than one of many millions. Since the jets develop over timescales proportional to the hole's size (and therefore to its mass), observing one of these nearby objects for a year may help us to understand what happens to a giant radio source over its lifetime of many million years.

Microquasars allow us to follow the jet production process in real time. Episodes of jet production are seen to develop over timescales of weeks to months. Often a fresh jet outburst is preceded by a powerful X-ray flare, as if the accretion disk were suddenly draining into the center. Microquasars also exhibit various systematic forms of variability on much shorter timescales, from minutes down to a few milliseconds. Some of the X-ray fluctuations consist of slightly irregular (*quasi-periodic*) oscillations that astronomers suspect contain clues to the gravitational field around the compact object as well as the physics of the accretion process. However, we are not yet sure how to interpret these oscillations (see Chapter 10).

Where should we place jets in the scheme of phenomena in the Universe? Their production is not restricted to black holes. They appear to be a rather general phenomenon, seen on all scales from single protostars to the span of a cluster of galaxies. Maybe they are apt to form anytime a flattened, rotating system manages to inject some extra energy into the material near the rotational axis. The concept of the magnetic propulsion of jets applies so widely that it too has acquired the cachet of universality, a quality that astrophysicists find almost irresistible. It seems to be the only plausible mechanism that can both accelerate protostellar jets, and propel the jets from active galaxies to more than 99 percent of the speed of light. But we still cannot rule out an important role for radiation pressure in some cases – for example, in SS 433 or in the jets of gamma-ray bursts.

The X-ray binary Cygnus X-1 has a one-sided jet that has inflated a lobe in the interstellar medium. Analysis of the lobe's energy content suggests that the energy output in the jet is at least as large as the radiation emitted by the accretion disk. The shock wave outlining the lobe can be traced in radio emission as well as optical emission lines of hydrogen and oxygen.

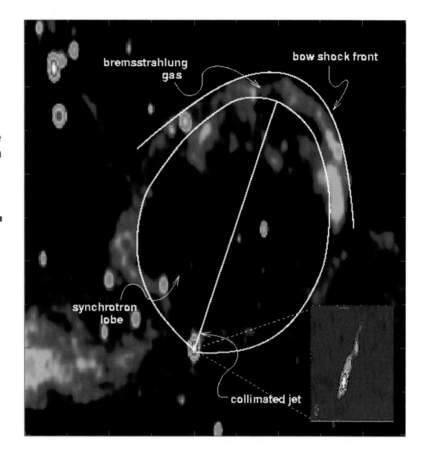

In the next chapter, we shall focus on gamma-ray bursts in more detail. In so doing, we are forced to confront the production of jets under even more extreme conditions than those that seem to apply in AGN or microquasars. The jets associated with gamma-ray bursts must have speeds within *a thousandth of a percent* of the speed of light, and carry off the energy equivalent of a fraction of a solar mass within a few seconds. Though there is much we do not understand about these flows, and what propels them, we will see that they are very likely signaling the births of stellar-mass black holes.

7

BLASTS FROM THE PAST

Astronomers are at a double disadvantage compared to other scientists. As in other sciences, we have to work hard to determine the physical nature of systems we are studying. But unlike a chemist in the lab, we are constrained to do this passively, by collecting observations from a remote vantage point. To make matters worse, we often don't even know where objects of interest are located. All we can see is a two-dimensional projection of their positions on the sky. Adding depth to that projection – finding the distances to objects – has required as much ingenuity and effort as any other endeavor in astronomy.

Thus, it took decades after the advent of photography before it was realized that galaxies are more than just spiral nebulae within the Milky Way. Until their true distances could be established, it was impossible for Edwin Hubble to discover the relationship between redshift and distance and to deduce the expansion of the Universe. Likewise, the immense distances to quasars were not accepted immediately by all astronomers, who questioned the applicability of Hubble's expansion law to objects so strange. Given these historical precedents, perhaps it is not too surprising that it took more than 20 years to discover the extraordinary nature of gamma-ray bursts.

The story of gamma-ray bursts dates back to the 1960s. American scientists at Los Alamos had developed the *Vela* system of satellites, whose purpose was to monitor clandestine tests of nuclear weapons in space by detecting the associated gamma-ray emission. Gamma rays are the form of electromagnetic radiation with the shortest wavelengths and, correspondingly, the highest energies – higher than X-rays. Occasional flashes were indeed recorded, but it took several years before the scientists became convinced that these gamma-ray bursts were natural rather than sinister in their origins, and weren't instrumental artifacts. Not until 1973 was the discovery announced in *The Astrophysical Journal*. Typical bursts lasted for several seconds, though there was often flickering on much shorter timescales.

This discovery stimulated an avalanche of theoretical speculation. Less than two years later, at the 1974 Texas Symposium on Relativistic Astrophysics, the American physicist Malvin Ruderman reviewed a long catalog of exotic ideas that had already appeared in the literature. He noted a tendency, still apparent in later work, for theorists to "twist rather strenuously to convince ourselves and others that observations of new phenomena fit into our chosen specialties." An updated list of gamma-ray burst theories, published in 1993, identified no fewer than 140 distinguishable models!

During the late 1970s and 1980s, satellites accumulated further data on bursts. The *Pioneer Venus Orbiter* (PVO) detected about 200 strong bursts over a thirteen-year period. The number of detected bursts rose faster than the number of proposed models – indeed, some

One of the 16 gamma-ray bursts reported in the discovery paper by Klebesadel *et al.* in 1973. The same burst is recorded by three different *Vela* satellites. Note the short spikes of gamma-ray emission superimposed on the overall burst, which lasts about 10 seconds.

of the wilder early conjectures were actually ruled out. Nevertheless, the observed properties of gamma-ray bursts (GRBs) were anything but "normal." The flux of the radiation from a typical burst – the amount of energy per square centimeter per second collected at the Earth – exceeds (albeit for only a few seconds) that of any known celestial object beyond the Solar System. They are never seen to repeat at the same location. There is no standardized spectrum or time profile of a burst, although gamma-ray bursts can be clearly distinguished from X-ray bursts, which are known to come from nuclear explosions on neutron stars accreting from binary companions, and from a rare class of objects known as "soft gamma-ray repeaters."

Between 1973 and 1991, most astrophysicists came to favor the view that gamma-ray bursts originate at neutron stars in our own Galaxy. The bursts were envisaged, for example, as random magnetic flares on the surfaces of old neutron stars whose pulsar activity had died out. These flares could be triggered by "starquakes" in the stars' rigid crust. Or perhaps neutron stars are surrounded by planetary systems (some examples are known!), and bursts signal the destruction of an asteroid or comet converting 10 percent of its mass-energy into radiation as it crashes into the star. However, not all theorists "bought" the old-pulsar hypotheses. The problem was that the distances to gamma-ray bursts were completely unknown, and positional measurements were too crude to tell whether the bursts coincided with known objects such as galaxies or star clusters. Some suggested that the bursts might originate as close by as the comet cloud on the outskirts of our own Solar System. Others supposed them to be at the edge of the Universe, perhaps the discharges of "superconducting strings" or other exotica.

Location, Location, Location

Even without directional information, one can learn something about how the bursts are distributed in space by comparing the relative numbers of strong and weak ones. If the bursts originate from uniformly distributed sources, then it is easy to calculate how the number of detected bursts would go up as the sensitivity of the detector improves. However, observations showed that the sources are clearly *not* uniform in space. Weak bursts are rarer, relative to intense bursts, than would be expected if the bursts came from a uniformly distributed population. It seems that the sources of bursts thin out beyond a certain distance. Unfortunately, this test alone does not tell us what that distance is. Could it be that the bursts do come from Galactic neutron stars, and we see the population declining with increasing distance from the Milky Way?

Plot showing the relative number of gamma-ray bursts brighter than a given level. There are fewer faint gamma-ray bursts than would be expected if they were distributed uniformly through space (dashed line), indicating that the burst sources "thin out" beyond a certain distance.

Positions of the 2704 gamma-ray bursts recorded by the Burst and Transient Source Experiment on the *Compton Gamma Ray Observatory* are plotted on the sky. As far as anyone can tell the directions are completely random. This favors the hypothesis that bursts come from the remote outposts of the observable Universe. If the bursts came from objects within our Galaxy, one would have expected a more evident concentration toward the plane of the Milky Way, or toward the Galactic Center.

2704 BATSE Gamma-Ray Bursts

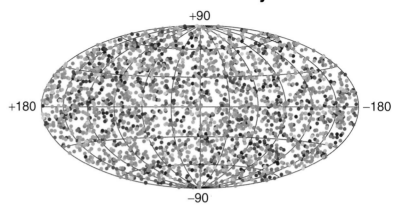

If we could see gamma-ray bursts from neutron stars all the way out to the edge of the Milky Way, then their distribution on the sky ought to reflect the asymmetric shape of the Milky Way as viewed from our vantage point – the fainter bursts, at least, should be concentrated in a band, just like the stars in the Milky Way. This was nearly everybody's expectation when, in 1991, NASA launched the *Compton Gamma Ray Observatory,* which carried an instrument for monitoring bursts. To nearly universal surprise, the nearly 3000 bursts detected during the 9-year mission seemed to come from random directions over the entire sky: there is no tendency for them to be concentrated toward the center of our Galaxy, nor toward the plane of the Milky Way.

This observation of *isotropy,* coupled with the relative brightness distribution, made it exceedingly difficult to explain the gamma-ray bursts as originating from *any* kind of object located in our Galaxy. Even a population of objects located in a halo around the Solar System would show detectable asymmetry, because the Earth is not located at the center of the Solar System. The same would be true for a hypothetical population of objects located in an extended halo around the Milky Way. Even if this halo were perfectly spherical, we would detect an asymmetry due to the Sun's 25 000 light-year offset from the Galactic Center.

The only known system that shows the required level of isotropy is the Universe itself – the distant galaxies are distributed around us with remarkable uniformity. This does not mean that we are located at some special "center" of the Universe. In fact, one of the strongest tenets of astronomers is that there is *nothing special* about our location. (This is an extension of Copernicus's assertion that the Earth does not lie at the center of the Solar System, and is therefore often referred to as the Copernican principle.) Theoretical models for the expanding universe, based on general relativity, have the property that the distribution of

Bohdan Paczyński argued persuasively that gamma-ray bursts were at cosmological distances, long before this idea was generally accepted. He also made fundamental contributions to the theory of stellar evolution and accretion disks, and initiated the search for dark matter based on gravitational "microlensing" (Chapter 4).

distant galaxies should appear isotropic when viewed from *any* location. Thus, the isotropy of gamma-ray bursts had, by the mid-1990s, led most astrophysicists to suspect that the bursts are a remote extragalactic phenomenon. The random distribution on the sky is then no problem, and if their distances are large enough – approaching the "Hubble radius" of the Universe – the redshift due to the expansion of the Universe would tend to diminish the number of fainter (and more distant) bursts detected (see the box on pages 80–81). Our inability to detect the fainter bursts would make it appear that we are seeing out to the "edge" of the distribution of burst sources.

Although the isotropy evidence had tilted the balance of opinion strongly toward the "cosmological" view, it still remained conceivable that the bursts were in our Galactic halo – provided that they were sufficiently far out that the Sun's offset did not produce a detectable asymmetry. In 1995 Bohdan Paczyński, one of the earliest and most influential converts to the "cosmological" view, debated Donald Lamb, a leading proponent of the neutron star hypothesis, at the National Museum of Natural History in Washington, D. C. The debate marked the 75th anniversary of the famous debate that took place there between Harlow Shapley and Heber Curtis on whether some of the so-called "nebulae" were stellar systems (i.e., galaxies) beyond our Milky Way.

There was general agreement on the kind of new evidence that could settle the issue of the distances to gamma-ray bursts. No redshift could be measured from the gamma rays themselves. Only by detecting the bursts in a different wavelength band, such as optical or radio, could one hope to obtain accurate positional or distance measurements. Unfortunately, each of the bursts detected by the *Compton Observatory* could be anywhere within a circle several degrees across – a patch of sky bigger than the field of view of most optical telescopes. A network of interplanetary probes, including spacecraft like the *PVO*, was able to do somewhat better for especially bright bursts – about an arc minute – by using a technique known as triangulation. This is still not good enough to fix whether (for instance) a burst originates in a particular galaxy, but it would suffice to tell optical astronomers where to point their telescopes in order to search for an "afterglow." However, triangulation was a laborious process; by the time a position had been calculated the burst had long faded.

It required only a few years to resolve the controversies of the Shapley–Curtis debate; those concerning the distances to gamma-ray bursts were settled even more quickly and decisively. The breakthrough came in 1997, after the launch of an Italian/Dutch satellite known as *Beppo–SAX* (the prefix was added to the original acronym in honor of the eminent Italian physicist Giuseppe [Beppo] Ochialini). This instrument was less sensitive than the *Compton Observatory*, and only detected about one burst per month. But *Beppo–SAX* had a crucial

The Italian–Dutch satellite *Beppo–SAX* discovered the first gamma-ray burst afterglow on February 28, 1997. These images show the afterglow in X-rays at 8 hours (left) and 3 days (right) after the burst.

advantage: it combined arc-minute positional accuracy with rapid response time. For the first time, positions of gamma-ray bursts could be made available within a few hours – an especially important advantage if the afterglow is shortlived. The first afterglow was discovered on February 28, 1997 and faded over the next few months. A second one, detected on May 8, nailed down the distance scale. Astronomers were able to take an optical spectrum with the Keck 10 m telescope in Hawaii. This spectrum displayed strong absorption features caused by gas with a redshift of 0.835. The burst was, therefore, either in a galaxy with that redshift, or was even further away, so that the light from its afterglow passed through this galaxy on its way to us. A burst in a galaxy at redshift 3.4 was discovered before the end of 1997, and subsequent observations have recorded bursts at redshifts up to 8 and higher. Typical gamma-ray bursts are even more distant than quasars – they are truly "blasts from the past."

Hypernovae and Collapsars

What kind of "trigger" could set off a gamma-ray burst? At such immense distances the typical *luminosity* emitted during the few seconds of a burst would have to be colossal, exceeding by many

(a)   (b)   (c)

The ultraluminous gamma-ray burst 990123 was located on the outskirts of a star-forming galaxy with a redshift of 1.6. The optical afterglow peaked at 9th magnitude, 45 seconds after the burst, and could have been seen with binoculars. This sequence of images from the *Hubble Space Telescope* shows the afterglow fading at (a) 16, (b) 59, and (c) 380 days after the burst.

thousands the most extreme output from any quasar. The total *energy*, however, is not out of line with some other phenomena encountered in astrophysics indeed, it is reminiscent of the energy released in the core of a supernova. As early as 1993 Stanford Woosley, of the University of California at Santa Cruz, suggested that a gamma-ray burst could erupt when the core of a massive star collapsed directly to a black hole (a "collapsar"). Some of the early afterglow measurements indicated that upwards of the rest mass energy of an entire solar mass – far more than a supernova's output – was being converted to radiation. Could gamma-ray bursts arise from some kind of extremely rare, extra-energetic form of stellar explosion – a "hypernova," as speculated by Bohdan Paczyński? The theoretical challenge posed by supernova models for bursts was to understand how the energy could all escape within seconds instead of being smothered by the stellar envelope, which slows and moderates the release of radiation in ordinary supernovae. Other models, such as the merger of two neutron stars or a neutron star and a black hole, seemed better able to cope with the short timescales.

Nevertheless, tantalizing clues favored the supernova hypothesis. Afterglows seemed to be located in star-forming regions of distant galaxies, suggesting an association with young stars. Neutron star binaries, on the other hand, take so long to spiral together that they should have had plenty of time to drift away from their spawning grounds. Propelled by a "kick" when the neutron stars first form, they should be seen mainly on the quiet outskirts of old galaxies. Then, in 1998, an unusual supernova was discovered very close to the position of GRB 980425 (meaning, the gamma-ray burst observed on April 25, 1998. Since bursts are seldom observed more than once per day, this nomenclature is adequate to identify them.) Supernova 1998bw, at the relatively low redshift of 0.0085, lay within the "error circle" for the gamma-ray burst, meaning that they *might* be the same object – although a chance alignment on the sky could

As the optical afterglow of GRB 030329 faded, the spectrum of the underlying supernova, SN 2003dh, became visible. The final spectrum (bottom), taken 34 days after the burst, resembles the spectrum of SN 1998bw at a similar age (dashed).

not be ruled out. It seemed like an odd association. SN 1998bw was much brighter than a typical supernova and its debris was expanding much more rapidly than normal. But if the burst was assumed to be at the same distance as the supernova, then it was much weaker – by several powers of ten – than a typical burst. Also, its gamma rays did not extend to the highest energies usually seen. Unfortunately, the burst was too weak to produce a detectable afterglow at visible wavelengths, which would have yielded a much more accurate position.

A more secure connection was established in 2003, when the spectral imprint of the light from supernova 2003dh was recorded on top of the fading optical afterglow of GRB 030329. This, and a few similar instances, convinced most astronomers that many gamma-ray bursts are indeed associated with stellar explosions. The dampening effect of the thick stellar envelope could be avoided if the collapsed core of the star produces a tightly focused jet, which bores through the envelope in a few seconds and keeps a channel open for subsequent energy release. The characteristic way in which the light of the afterglow fades – slowly at first, then faster – was interpreted as evidence for a collimated flow that gradually slows down and eventually spreads out sideways.

This schematic diagram illustrates the collapsar model for gamma-ray bursts and their afterglows. The left image shows a pair of jets (yellow) being launched from an accretion disk (red) that is fed by matter collapsing into the center of the stellar envelope (light blue). In the central image, the jets have broken through the stellar surface and emit a "prompt" burst of gamma rays. Later, the jets slow down and spread out as they plow into surrounding matter, producing the afterglow (right image).

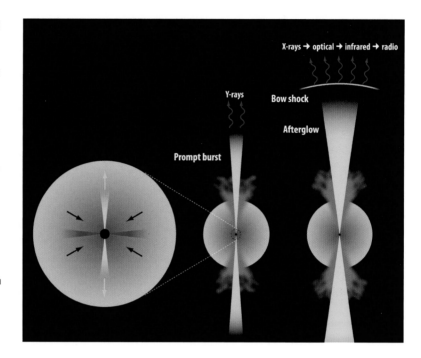

The deduction that gamma-ray bursts are emitted by jets, and can only be seen from a narrow range of directions, solved a puzzle about their energetics. The most luminous bursts, it seemed, put out 100 times more energy than a supernova. But this number was calculated by assuming that the burst could be seen from any direction. If all the gamma rays were focused into a narrow cone covering, say, 1 percent of the sky (as viewed from the collapsar), then the energy requirements would be reduced accordingly.

Although especially energetic "hypernovae" are no longer needed to explain gamma-ray bursts, this does not mean that *every* supernova must be capable of producing one. Approximately one burst is observed per day, but if we could detect every supernova that occurred within the Hubble radius their frequency would amount to one per *second*. If beaming renders 99 percent of bursts undetectable from Earth, this still means that only one in 1000 supernovae has to produce jets in order to explain all the bursts we see. Like SN 1998bw, the supernovae associated with gamma-ray bursts do seem to be unusual – more luminous than typical supernovae and with larger expansion velocities.

Does this mean that every gamma-ray burst is triggered by the collapse of a massive stellar core? Probably not. Most astronomers believe that supernovae account for the long-duration gamma-ray bursts, those lasting more than a couple of seconds. But about a third of

This histogram shows that there are two classes of gamma-ray bursts, "short" bursts with durations of about 2 seconds or less, and "long" bursts lasting many seconds.

all bursts fall into a distinct category – short bursts – that last less than 2 seconds, often much less. Since even a tightly collimated jet cannot penetrate a stellar envelope in less than a couple of seconds, supernovae were not considered good candidates to provide the short bursts. Confirmation of this fact was long in coming, however. No short burst yielded a detectable afterglow until the *Swift* satellite was launched in 2004. In addition to a gamma-ray burst detector, *Swift* carried sensitive X-ray and ultraviolet/optical telescopes that could localize a burst and begin monitoring its afterglow within 100 seconds.

The data from *Swift* showed that short bursts and long bursts occur in completely different environments, and are almost certainly triggered by different kinds of events. Short bursts occur on the outskirts of elliptical galaxies, far from any star-forming region. They are less luminous than long bursts, on average, and their host galaxies tend to be closer to us. The most likely trigger involves a pair of neutron stars (or maybe a black hole and a neutron star) in a close binary orbit. The two stars spiral together, slowly at first and then faster and faster, until they finally merge. The final coalescence, lasting only a millisecond, would certainly trigger a large enough release of energy, and the subsequent draining of debris from the merger could easily draw out the energy release over a second or more. Such events are rare – they would occur only once every 100 thousand years in a typical galaxy. But there are more than 100 billion galaxies within the Hubble radius, so the rate at which short bursts are detected poses no real problem. Reassuringly, no underlying supernova signatures have been found to be coincident with short burst afterglows.

Modeling Gamma-ray Bursts

The identification of long gamma-ray bursts with supernovae seems relatively secure. But our understanding of how these outbursts are ejected from the stellar core, and how they convert their energy into gamma rays, lags far behind even our incomplete understanding of the jets in AGN. Our understanding of the short gamma-ray bursts is even more primitive.

What we can say is that the energy responsible for the initial burst of gamma rays – the "prompt" emission – in both long and short bursts is propelled toward us at an extremely relativistic speed. Without relativistic motion, the short timescales on which the gamma rays fluctuate, as short as milliseconds for both kinds of bursts, would imply an emitting region so compact that the gamma rays could not escape. Instead, the highest energy gamma rays would be converted into such a dense soup of electrons and their corresponding antiparticles, positrons, that it would trap the rest of the photons as well. As we saw in Chapter 6, when we considered the escape of gamma rays from AGN, relativistic

Gamma-ray bursts often fluctuate wildly on very short timescales, as shown in these light curves of three bursts. These very rapid fluctuations indicate that the matter producing the gamma rays is moving toward us at very close to the speed of light.

motion makes rapid variability compatible with a large emitting volume. In the case of gamma-ray bursts the requirements are extreme: the speed has to be within a few thousandths of a percent of the speed of light (compared to a fraction of a percent for a typical AGN).

How is gas accelerated to such extreme speeds? Astronomers have no firm leads, but are pursuing the "usual suspects" – magnetic fields, tapping the spin energy of either an accretion disk or a black hole – as well as more exotic effects, such as intense heating by neutrinos or acceleration by trapped radiation. A merger between two neutron stars, or a neutron star and a black hole, should produce a rapidly spinning black hole surrounded by a thick, dense disk of gas. The collapse of a massive stellar core, if it starts out with sufficient rotation, probably leads to a similar situation. In both cases, twisting and compressional motions could amplify the strength of the magnetic field to levels capable of propelling the outburst. A strong magnetic field could, of course, provide the initial focusing of the outburst into a collimated jet, but the whirlpool-like funnel of the thick disk could also play this role.

Little is known about how a black hole/disk/jet system would evolve under the conditions appropriate to gamma-ray bursts. Much would depend on the distribution of angular momentum in the collapsing core, or the details of the merger. Some of the gas ejected during the initial collapse or merger would not have enough energy to escape and would continue to rain down onto the disk, extending the duration of the burst and giving its output a complex temporal structure. Indeed, the sensitive X-ray observations made by *Swift* suggest that a burst's central engine can last far longer than expected, stopping and restarting multiple times.

One of the most surprising aspects of the collapsar model for long bursts is that a highly relativistic jet punches through a stellar envelope without slowing down or losing its integrity. Numerical simulations have been valuable in demonstrating the plausibility of this idea. Like the jet from a radio galaxy boring through the intergalactic gas, a collapsar jet insulates itself with a cocoon of pressurized, hot material as it traverses the envelope. Even if the jet starts out weakly focused as it

This computer simulation, by Maximilian Ruffert and Hans-Thomas Janka, shows a neutron star binary in the process of merging. Mergers between neutron stars (or a neutron star and a black hole) are promising candidates for short gamma-ray bursts. In the bottom frames, it has become a single object with a disk around it; subsequently it collapses to a black hole. A considerable amount of energy in gravitational radiation can be released during the intermediate stages of the merger; we discuss the prospects for detecting this radiation in Chapter 10.

leaves the central engine, the high pressure of its surroundings collimates it tightly as it traverses the envelope. As a result, the jet can make its way to the stellar surface within a few seconds. As it exits, the much slower cocoon gas begins to exhaust through the hole, perhaps providing an explanation for the X-ray "precursors" to gamma-ray bursts that are sometimes observed.

This computer simulation, by Brian Morsony and Davide Lazzati, shows an ultrarelativistic jet penetrating a star and emerging through its surface. Panel (a) shows the jet just exiting the star after 7.75 seconds, while panel (b) (which represents a region 50 times larger) shows the jet having traveled far beyond the stellar surface at 87 seconds after launch. The upper half of each panel shows the density of matter, on a scale ranging from high density (red) to low density (dark blue), while the bottom half shows how close the speed is to the speed of light; here, red represents a slightly slower speed. Note how the jet gradually widens with time as it disrupts the star, but still maintains its high speed and focused trajectory.

(a)

(b)

This diagram illustrates how the gradual widening of a collapsar jet could produce an X-ray "precursor" to a gamma-ray burst. In the left image, the jet has only recently emerged from the star and is very narrow. Viewing from slightly off-axis, the observer sees the X-ray-emitting debris from the stellar interior but cannot see the gamma rays emitted by the jet, which are strongly beamed in the jet direction (Chapter 6). Later, the jet has widened to the point where the observer can see the gamma rays (right image).

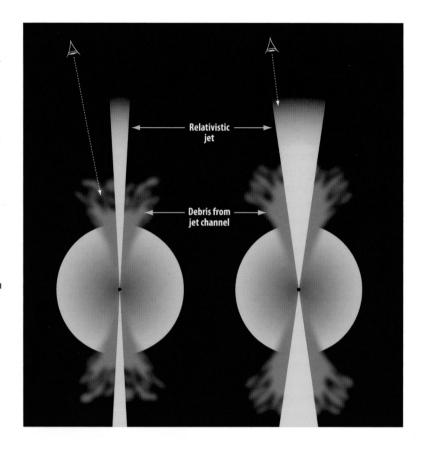

As continuous flow of the jet keeps the channel open, the jet's structure is bound to evolve. High pressure pushes the stellar gas aside, widening the channel and, presumably, the jet. An observer, not initially in the jet's Doppler beam, might see a delayed start to the burst as the jet expands into his line of sight. While *Swift* data show a complex phenomenology that could be associated with factors like viewing direction and jet evolution, we should be cautious about overinterpreting the observations. After all, the jet model for gamma-ray bursts was initially inspired by observations of a simultaneous steepening in the rate at which the afterglow fades at different wavelengths. With the more sensitive measurements of the *Swift* era, these "achromatic breaks" are increasingly viewed as observational artifacts. Despite the lack of direct evidence, though, no one is ready to give up the idea that gamma-ray bursts are focused into jets – it provides too neat an explanation for other aspects of gamma-ray burst behavior.

It is especially difficult to understand how so much of the energy can come out in the form of gamma rays. The collapsed core of a massive star, or the debris disk left over from a neutron star merger, is so dense and opaque that the "prompt" escape of radiation is impossible. The radiation would be trapped, and its energy converted into kinetic form. In other words, the relativistic flow required to avoid the production of too many electrons and positrons must also carry most of the energy released in the burst. To produce gamma rays, this energy of motion must be converted back into energy of radiation. But how? Theorists have suggested several possible solutions. One idea is that the jet is very unsteady, with many internal velocity fluctuations. When a faster region catches up to a slower region, a pair of shock waves is generated and the relative kinetic energy is dissipated into radiation. But this mechanism can work only if magnetic fields are energetically unimportant in the flow – otherwise the gas would be too "springy" to generate shocks. Thus, there are other models in which the radiation is produced via the dissipation of magnetic energy. Presumably, once we have a clearer idea of how the energy is dissipated, we might also understand which mechanisms actually generate the photons. The spectrum of a gamma-ray burst "looks" nonthermal, although there is usually a broad peak in the emitted energy at wavelengths corresponding to low-energy gamma rays. Most researchers have assumed that the radiation is produced by the synchrotron mechanism, although there is no direct evidence for this. Other mechanisms are certainly viable; it is even possible that the broad spectral peak is a relic of thermal radiation generated deep in the jet.

When a supernova explodes, the expanding remnant sweeps up external matter into a hot shell. The debris from a gamma-ray burst would likewise explode into the surrounding medium. The burst ejecta create a blast wave that initially moves at nearly the speed of light, but slows down as it runs into external material. The radiation from this swept-up material is the afterglow; studying its properties should tell us something about the burst's environment as well as the power and geometry of the jet. To some extent observations have borne out the main theoretical predictions for the evolution of the afterglow: a gradual decrease in brightness, lasting months, with a smooth transition from X-rays to optical to radio emission. But *Swift* has shown us that the real situation is far more complicated than the idealized models suggest. We have already noted that the "achromatic break" in the light curve that was supposed to signify a jet has gone by the wayside. The optical stage of the afterglow is often very faint or undetected, and the track of the X-ray intensity is now seen to be so complicated that it is impossible to know where the prompt emission ends and the afterglow begins.

Thus, gamma-ray bursts remain something of a mystery. From their energetics, locations and timescales we deduce that they represent yet another context in which black-hole stellar remnants or neutron stars reveal their presence with displays of cosmic fireworks. Do the long bursts signal the formation of stellar-mass black holes? We still don't know. If they are beamed, as we suspect they are, then their energies are just about compatible with those of ordinary supernovae. If the collapsar engine does involve very rapid accretion (at rates exceeding 0.1 solar masses per second) over tens of seconds, then it is likely that the central object ultimately collapses to a black hole, even if it starts out as a neutron star. But all the energy for a collapsar burst could instead be supplied by the rapid spin and super-strong magnetic field of a neutron star. Even less can be said about the outcome of the short bursts. Whatever the central engine, the flow speeds emanating from gamma-ray bursts are extraordinary. In Chapter 6 we speculated that the somewhat less relativistic speeds of AGN jets might be the "smoking gun" of energy extraction from a spinning black hole. But we also pointed out that this interpretation has been called into question by evidence that neutron stars can also produce highly relativistic jets.

Why has it proven so hard to pin down the nature of gamma-ray bursts? We attribute the difficulty to three factors. First, the bursts are very short-lived and erratic phenomena. One has to be nimble to catch both the dying of the prompt emission and the onset of the afterglow. *Swift* has shown us that it will be no simple task to disentangle these two, given the hiccoughs and sneezes that seem to characterize the X-ray light curve. Second, bursts are both small and distant. We can trace the jets of the most distant quasars down to scales of a few light-years, using radio interferometry. By the time the putative jets of gamma-ray bursts reach these scales, they have probably lost their collimation and may have faded into the background. Finally, gamma-ray bursts lack the transparency of at least some AGN. With the dense debris cloud that surrounds a burst, there is at least a factor of 10 000 in scale between the size of the central engine and the radius from which gamma rays begin to stream in our direction. We therefore leave gamma-ray bursts with a slight feeling of pessimism and turn to black holes that are both closer to home and much less dramatic in their manifestations. For the next two chapters we once again consider the supermassive black holes at the centers of galaxies, this time treating them not as spectacular oddities but as essential elements of galactic structure and evolution.

8 BLACK HOLES IN HIBERNATION

If you select a nearby galaxy at random and observe its nucleus, there is only one chance in a hundred that you will detect moderately vigorous activity attributable to a massive black hole. The chance of stumbling on a powerfully active galaxy – a quasar or giant double radio source – is even smaller. But this does not mean that massive black holes occur only in the nuclei of a small fraction of galaxies, for activity requires not only a massive black hole but also a supply of "fuel." To power a quasar or luminous Seyfert nucleus by accretion there must be more than a solar mass per year of gas flowing into the black hole. Radio jets powered by a black hole flywheel might survive on a smaller yearly mass budget, but a substantial gas supply is nevertheless needed to maintain the magnetic fields in the hole's vicinity. And, without more vigorous episodes of accretion, the hole's rate of spin will slowly decline.

A solar mass of gas per year is a sizable fuel supply, even by galactic standards. It is unlikely that the nucleus of a galaxy maintains this level of gas inflow for more than a small fraction of its lifetime. Perhaps there is a time early in the galaxy's life when the stars and gas in the nucleus become so closely packed that some kind of runaway catastrophe occurs, dumping matter copiously into the black hole. From time to time, the galaxy might merge with a smaller, gas-rich galaxy or swallow an intergalactic gas cloud – we see examples of such encounters. For a while following such an event, matter might pour into the nucleus at a rate fast enough to power a quasar. But such episodes of violent activity most likely represent a short-lived phase in the life of a galaxy. When the fuel supply dwindles, the activity dies away but the black hole does not disappear. Black holes in "hibernation" – massive black holes now starved of fuel, and therefore quiescent – apparently lurk at the centers of virtually all large galaxies, including our own Milky Way. Their masses are correlated with properties of their hosts, indicating a link with the epoch of galaxy formation in the distant past. These objects can provide clues to the enigma of quasars, and remind us that even the benign-looking galaxies around us could reawaken and turn violent once again.

The Milky Way's Center, Revisited

The richness of the star-field at the center of our Galaxy becomes apparent only at infrared wavelengths, which can both penetrate the intervening dust and reveal a significant fraction of the stellar radiation. Increasingly sensitive infrared detectors, developed during the 1970s, 1980s and 1990s, initially were restricted to sensing the cumulative light of star clusters. But later they improved enough to record the locations and spectra of individual stars. These observations revealed a large number

Infrared images of the central region
of the Milky Way show a dense
concentration of stars. This image
covers about 4 light-years on a side;
the central black hole is located at the
center of the small rectangle.

Reinhard Genzel, from the Max Planck Institute for Extraterrestrial Physics in Garching, Germany, is an innovative pioneer of instrumentation for making observations in the infrared band. By observing in this band (where the obscuration due to interstellar dust is less than it is for visible light), both he and Andrea Ghez were able to study the motions of stars very close to the Galactic Center. They discovered that some were very fast-moving, and were tracing out elliptical orbits around a central compact mass – almost certainly a black hole.

of hot, bright stars that must have formed relatively recently – within the last few million years – in addition to a population of older stars. Unlike the stars near the Sun, these stars do not participate in organized rotation, but have random velocities averaging about 90 kilometers per second. From these velocity measurements, plus the assumption that the stars are held in bound orbits, one can estimate the amount the mass preventing the stars from flying apart, in much the same way that Fritz Zwicky deduced the existence of dark matter in the Coma Cluster. Using the European Southern Observatory's 3.5 m New Technology Telescope (NTT) at La Silla, Chile, the group led by Reinhard Genzel at the Max Planck Institute for Extraterrestrial Physics near Munich, Germany, confirmed what earlier measurements had suggested: that the typical stellar velocities were inching up, the closer one got to the strange radio source at the exact center of the Milky Way, Sgr A* (Chapter 4). The mass needed to hold the stars in place was about 3 million solar masses.

As we noted in Chapter 4, stars are observed to be roughly 300 times closer together in the Galactic Center than they are in the Sun's neighborhood, implying that more than 100 000 of them occupy each cubic light-year. At this density, it would require a volume of 30 cubic light-years to enclose 3 million solar masses. But Genzel's result indicated a far higher concentration of mass – the 3 million solar masses had to be contained within a volume of only 0.03 cubic light-years. If the velocities continued to creep up at smaller distances from Sgr A*, it would imply even more extreme conditions for the central mass. But Genzel's group faced a technical problem. Not only were their measurements approaching the limiting angular resolution of the NTT, but also their observations were limited to short exposures on those rare occasions when the atmosphere was especially stable. Most of the time, turbulence in the atmosphere made the stars of the Galactic Center twinkle, blurring their images and slightly shifting their positions. Since stars were plastered so closely together on the patch of sky surrounding Sgr A*, this created a "confusion" problem: one could not be sure that one was taking a spectrum of a single star, or a merged image of two or more. (Going outside the atmosphere was not an option: the *Hubble Space Telescope* had a smaller aperture than the NTT, and thus lower resolution; and it was not optimized for observations at infrared wavelengths.)

Genzel's group, and a group led by Andrea Ghez at the University of California, Los Angeles (using the 10 m Keck Telescope in Hawaii), solved this problem by taking thousands of extremely short exposures, lining them up, and stacking them together to mimic a single long exposure with most of the effects of atmospheric turbulence removed. By repeating the process every few months, they could track changes in the positions of stars on the sky, and they discovered one star that seemed to be moving extraordinarily fast: about 1400 kilometers

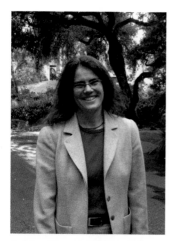

Andrea Ghez, a professor at the University of California, Los Angeles, is best known for her studies of the stars around the Galactic Center, carried out using special imaging techniques on the Keck Telescopes in Hawaii. She mapped out the orbits of stars close to the central black hole, gathering data complementary to the results obtained by Reinhard Genzel with telescopes in Chile.

per second. If this were its orbital speed, it would imply that the 3 million solar masses had to be crammed into a region with a radius less than a hundred times that of Neptune's orbit!

Genzel and Ghez knew that the short-exposure procedure they had mastered in the mid-1990s was a stop-gap measure. A much more powerful method to compensate for atmospheric turbulence, called "adaptive optics," had already been devised, although its implementation had to wait for increases in computer speed. In adaptive optics, the distorted and flickering image of a star is compared – every few milliseconds – to the stationary, pointlike appearance it would have in the absence of turbulence. Working backwards, one can then correct for the turbulence in real time by slightly deforming the shape of the telescope's mirror. The same correction not only stabilizes the image of the star being monitored, but also that of any object sufficiently near it on the sky. Stars bright enough to be measured within a few milliseconds are relatively rare, but this has not hampered the use of adaptive optics on the world's largest telescopes, including the Keck Telescopes and the European Southern Observatory's Very Large Telescope (VLT), successor to the NTT. One does not need to use a real star to monitor atmospheric turbulence – a laser beam at a carefully chosen wavelength (often, the yellow light of sodium) is reflected by the upper atmosphere, producing an artificial star that works nearly as well.

Adaptive optics has dramatically improved the performance of large telescopes observing at infrared wavelengths. It has allowed the Ghez and Genzel groups to deduce the complete orbits of more than a dozen stars in the Galactic Center. By 2005 there was no doubt that the stars orbited a single compact mass located very close to the position of Sgr A* – and very little possibility that the mass could be anything other than a black hole. One star was seen to swing around the center of mass at a speed of 12 000 kilometers per second, placing its point of closest approach at less than 1000 Schwarzschild radii from the event horizon.

The center of the Milky Way has yielded the most compelling evidence to date for the existence of a supermassive black hole. But it is by no means the only compelling case. Next we will see how a unique set of radio measurements have revealed a thin orbiting disk of gas in the center of the nearby galaxy NCG 4258. The motions of the gas provide nearly as incontrovertible a case for a supermassive black hole, and an even more precise estimate of the black hole's mass.

Clues from Orbiting Gas

Unlike the Milky Way, the spiral galaxy NGC 4258 is often classified as active; its nucleus contains an intensely hot emitter of X-rays and it has

By measuring how the velocities of stars and gas change with distance from Sagittarius A*, the Genzel and Ghez groups built a very strong case for a massive black hole. Far from Sgr A* the velocities vary as if the stars and gas clouds themselves are providing the gravity. But close to Sgr A* the stars and gas behave as if orbiting in the gravitational field of a fixed mass of about 3 million suns (blue curve). If the gravity were supplied by a compact cluster of stars instead of a black hole, the cluster would have to be implausibly dense, as shown by the yellow, purple and green curves which illustrate various cluster models.

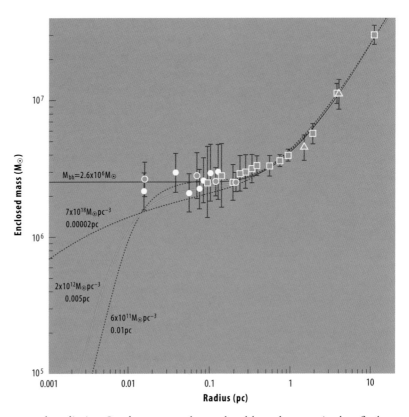

a weak radio jet. On these grounds we should not be surprised to find that the nucleus of NGC 4258 contains a supermassive black hole. But the evidence that clinches the case for a black hole in this galaxy, and allows us to measure its mass and distance accurately, comes from something much cooler – molecules of water in gas clouds orbiting very near the nucleus. The water molecules emit powerful microwave radiation because they act collectively as a maser.

The term "maser" is an acronym, standing for "microwave amplification by stimulated emission of radiation" (the acronym "laser" is the same, except that "light" replaces "microwave"). The microwave radiation in a maser is amplified to form an intense beam. The origin of the amplification lies in a sort of "domino effect." Spectral emission lines are produced when an atom or molecule is placed into an "excited state" – that is, a state of higher energy. Left alone, the excited atom or molecule will drop back into a lower-energy state, emitting a photon with a very specific wavelength. A series of these photons of specific wavelength forms the radiation that creates the spectral emission line. If you have a group of molecules in the same excited state, it is impossible to predict when any one of them will emit. Therefore, this "spontaneous emission" leads to photons emerging at random, or *incoherently*. But

184

© Adam R. Contos, 2002

Large telescopes, such as the 10-meter Keck Telescopes in Hawaii, use adaptive optics and laser "guide stars" to correct for the blurring due to the atmosphere.

Adaptive optics with a laser guide star works best at near-infrared wavelengths. This diagram shows the dramatic improvement in the ability of the Keck Telescopes to discern individual stars in the Galactic Center, when the adaptive optics system is switched on.

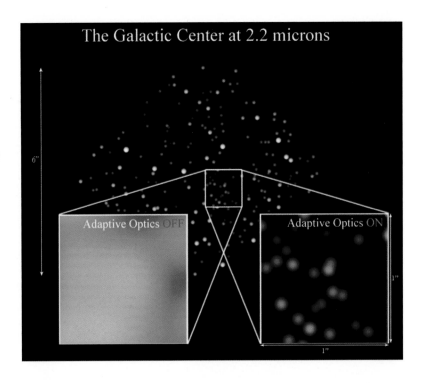

Painstaking observations over more than a decade have enabled the Ghez and Genzel groups to trace the orbits of individual stars around Sgr A*, providing incontrovertible evidence for a black hole.

When a photon hits an atom or molecule that is in an excited state, the photon can de-excite it, triggering (or "stimulating") the emission of another photon moving in the same direction. This amplification process can lead to very intense radiation.

there is a way to induce the molecules to emit sooner, and more predict-ably. If an intense beam of radiation with exactly the same wavelength as the spectral line passes by, it can "stimulate" a molecule to emit right away and add its photon to the beam, amplifying it. This effect becomes dramatic if you start with an anomalously large number of molecules in the excited state. Then the whole group of molecules can behave like a row of dominoes, all falling in step, and the result is a concentrated beam of radiation at uniform wavelength.

Masers and lasers require finely tuned conditions in order to operate in the laboratory (or in your CD player, or in any number of other items in everyday use). Therefore, astronomers were surprised to find that a similar process of "masing" occurred in the atmospheres of cool stars and in many interstellar clouds in our own Galaxy, leading to intense "hot spots" of emission in the spectral lines of various kinds of molecules. Exactly how the effect is triggered in astrophysical situations is uncertain. The essential point, however, is that the high intensity of the maser radia-tion enables radio telescopes to detect individual gas clouds in the galaxy NGC 4258; moreover, the water molecules emit a sharp spectral line with known frequency (22 GHz, corresponding to a wavelength of 1.36 centi-meters), so that the Doppler effect can easily be measured.

Very-long-baseline interferometers yield images of the molecu-lar clouds with angular resolution better than half a milliarcsecond. Moreover, the orbital speeds (about 1000 kilometers per second) can be measured with an accuracy of one part in 1000. The masing clouds delineate a slightly warped orbiting disk of gas, which we are viewing almost edge-on. The motions in this disk suggest that it is orbiting a central mass of 36 million suns. How do we know that these motions are gravitational in origin? Because so many individual clouds have been detected that we can actually plot a graph of orbital velocity vs. distance. The results show almost perfect agreement (to within about 1 percent accuracy) with the predicted velocity curve for objects orbit-ing a centrally condensed mass. This stunning agreement implies that virtually all the central mass must lie within the orbital radius of the *innermost* cloud we can detect. If more than even 1 percent of it spilled over the inner cloud radius, the fit would be spoiled. If we supposed that

By using the Very Long Baseline Array, radio astronomers have pinpointed the positions of the maser spots in NGC 4258 to better than a thousandth of an arcsecond (a milliarcsecond, or mas). The maser spots are clumped into three groups, which outline the foreground and extremities of a slightly warped disk, shown in the artist's impression below. Measurements of the Doppler shift reveal that the disk is rotating in the sense indicated by the color-coding of the dots. The entire disk is a hundred times smaller, in angular size, than anything that could be resolved with existing optical instruments, including the *Hubble Space Telescope.*

The center of NGC 4258 harbors a thin disk, shown here in an artist's impression, which we observe almost edge-on; the graph shows the spectrum of intensity vs. velocity, calculated from the Doppler shift. The disk contains cool molecular gas, which does not emit uniformly. Radio astronomers detect many discrete patches of emission, each coming from a region where the water molecules are acting like a maser.

The wavelength of the emission from each maser spot in NGC 4258 can be determined so accurately that the velocity can be inferred (from the Doppler shift) to a precision of 1 kilometer per second. The galaxy itself is moving away from us at 500 kilometers per second. When plotted against distance from the center, the velocities trace out a very clear pattern. The continuous line shows what Newton's theory would predict if the gas in the disk were orbiting under the influence of a 36-million-solar-mass central body. The disk has a sharply defined inner edge, at which the orbital speed is 1080 kilometers per second. Outside the edge, the orbital velocity falls off as the square root of the radius, exactly as expected if the gas is moving under the influence of a centrally concentrated mass.

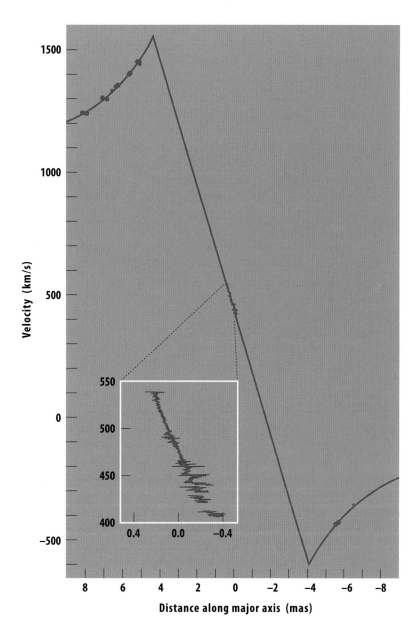

this mass were uniformly distributed inside the inner cloud radius, and dropped to zero sharply outside, its density would have to exceed 100 million solar masses per cubic light-year. This is already far higher than any density ever detected in a galactic nucleus or a globular cluster. But a real star cluster could never have such a sharply defined boundary. Fitting a realistic model of a star cluster to the velocity curve would require a central density *1000 times higher*. This approaches the limits on the

Milky Way's central density that were available in the late 1990s, but does not reach the standard set by recent determinations of individual stellar orbits. Nevertheless, such high densities and a total mass of 36 million suns cannot be accommodated in a cluster consisting of any known types of objects. The only plausible alternative is a single massive black hole.

So far, only the Milky Way and NGC 4258 have yielded up such clear evidence for massive black holes. This is not surprising for our Galaxy, since the stellar trajectories that have been mapped near the center would be too hard to discern in a distant galactic nucleus. The motions of masers, however, could be measured at much larger distances than the 24 million light-years that separate us from NGC 4258. But masers are unlikely to provide a general method for discovering black holes. Maser emission travels in narrowly defined directions, that depend on the velocities of the gas clouds and their geometric relationships. In all probability we see the masers in NGC 4258 precisely because the gas motions are so well-ordered and because the disk is so very nearly edge-on. Other AGN show strong maser emission, and some exhibit the signatures of orbital motion and edge-on geometry, so we may get lucky again. But it will remain the rare black hole that can be detected and measured in this way.

How much evidence is there, then, that other "quiet" galaxies host supermassive black holes in their nuclei? The answer is, "plenty." Although the evidence is not as clearcut as it is for NGC 4258 or the Milky Way, we can state confidently that most large galaxies contain a central massive black hole. Surveys have already probed the demographics of these holes, and reveal a remarkable relationship between their masses and the properties of their hosts.

A Black Hole Census

A truly quiet black hole can only be detected through its gravitational influence on its surroundings. In our attempt to observe such holes, the lack of activity can work to our advantage; the fireworks in an AGN often make it *more* difficult to detect the black hole "cleanly," that is, through its gravitational effects. AGN contain fast, even relativistic gas. But as likely as not, these motions are the result of nongravitational forces such as those due to radiation and magnetic fields. There are motions that obviously would be dominated by gravity, such as stellar orbits or the motion of gas in an accretion disk, but their signals are often drowned out by the intense radiation from the AGN.

We've seen that stellar-mass black holes can be detected cleanly in X-ray binaries. Measuring the Doppler shift of spectral lines from the normal star allows one to deduce a lower limit to the mass of the compact star in the binary. If this mass turns out to be larger than about 3 solar masses, one can comfortably conclude (by a process of elimination) that the compact star is a black hole. We would like to make the same kind of argument for a galactic nucleus. However, unlike an X-ray binary, where only two bodies are orbiting each other, in a galactic nucleus there are millions of stars in orbits whose properties (orientation, orbit radius, velocity, etc.) differ from star to star. We can only measure an average of the stellar orbits, projected through the star cluster. To make matters worse, the angular resolution is never adequate to isolate the region very near the black hole, so we must infer the hole's presence from trends observed on much larger scales. Nevertheless, several methods of analysis have led to convincing evidence for the existence of massive black holes in dozens of quiescent galactic nuclei.

Stars in a galactic nucleus move in random directions, like stars in a globular cluster. Each star in a galaxy responds to the sum-total gravitational force from all the other stars, as well as from any other matter that happens to be present (such as a black hole). The stars swarm around with a mean speed v such that the disruptive effect of kinetic energy balances the tendency of gravity to pull them all together at the center. This average speed can be different at different distances from the center, but its variation with radius follows distinct patterns in galaxies with and without black holes. In a galaxy without a central massive black hole, this speed levels off toward the center at no more than a few hundred kilometers per second, and may even decline for stars that spend most of their time near the very middle. But a black hole changes things markedly. Stars that venture within a certain distance of the hole move much faster in response to the extra gravitational pull. Thus v would increase toward the center.

The radius of the black hole's "sphere of influence" in the nucleus is determined by how close a star has to get before the hole's gravity overwhelms that of all the other stars combined. This is just the radius within which the escape velocity from the black hole's own gravity (or, almost equivalently, the orbital speed of stars gravitationally bound to it) becomes a large fraction of v. This sphere of influence has a radius proportional to the hole's mass and inversely proportional to the square of v. It is millions of times larger than the size of the hole's horizon, but it only subtends an angle on the sky of, at most, a few arcseconds, even in the nearest galaxies.

The extra gravity contributed by the black hole alters the motions of stars in a second, more subtle way. It increases the number of stars found close to the black hole at any one time, relative to the number that would be present if the black hole were absent. This extra

A massive black hole in the center of a galaxy has a "sphere of influence" within which the hole's gravity severely affects the motions of stars. A star entering the sphere of influence will be accelerated and deflected toward the hole. Other stars may be trapped on elliptical orbits confined within the sphere of influence. The black hole would therefore signal its presence by causing a concentration of fast-moving stars around it. A star on an almost radial orbit may get so close to the hole that it is swallowed or torn apart.

concentration becomes pronounced as the sphere of influence grows to encompass the orbits of more and more stars. Such a "cusp" in the stellar distribution can be observed as a sharp enhancement in the starlight at the very center of the galaxy. If the black hole has been present long enough for the stars in the cusp to interact gravitationally with each other on a one-by-one basis (which might take as long as a billion years), then the cusp can deepen still further as more stars become trapped within the sphere of influence.

Thus two telltale signs in the population of randomly moving stars could give away the presence of a quiescent black hole. The first is a central "blip" in the starlight, signaling the enhanced concentration of stars within the hole's sphere of influence. Such "blips" have been searched for, and found, but there is always an ambiguity in that the non-stellar light from a weak AGN – which might result from accretion onto a central black hole of relatively low mass – would be hard to untangle from the starlight of the cusp. A second, less ambiguous, signature would be evidence that the stars nearest the center are moving anomalously fast. Of course, ordinary stars cannot be detected individually in these galaxies. The sharpest detail that a large ground-based telescope can resolve, using adaptive optics, corresponds to a region more than a tenth of a light-year across at the distance of a nearby galaxy such as Andromeda, and such a region contains large numbers of stars. The *Hubble Space Telescope* cannot even do this well. Nevertheless, one can take spectra of the combined light from such a region and find lines characteristic of stellar spectra. The redshift and breadth of these lines tell us about both the organized and random motions of the stars, much as the broad emission lines tell us about the motions of gas clouds in AGN (Chapter 5).

Nearby galaxies obviously offer the best prospects for discerning the signatures of a quiescent black hole. The giant elliptical galaxy M87 is a radio source with an unusual optical jet (the one photographed by Heber Curtis in 1917) as well as weak extended radio lobes. The presence of these features indicates some central activity, though at a power level far below that of quasars. M87 is therefore a good candidate for having a central black hole in relative quiescence. In 1978, Peter Young, Wallace Sargent, and their collaborators at the California Institute of Technology claimed to have found evidence for a mass of 3 billion suns in its center. This claim was made on both of the grounds described above: a central peak in the starlight was attributed to the enhanced concentration of stars gathering around a central dark mass; and the typical random velocities appeared to rise from 300 kilometers per second to more than 500 kilometers per second as one observed on smaller and smaller scales, as would be expected if a central mass were present.

The credibility of this claim was debated throughout the 1980s. Part of the central peak in the light is now believed to originate not from stars but from a low level of accretion onto the central black hole. In addition, interpreting the random velocities has turned out to be trickier than expected. The problem is that our information about the average stellar velocity is incomplete. When we infer the random velocity by measuring the Doppler width of a spectral line, we are only detecting the part of the velocity that is along our line of sight. But the typical random velocity of a star may be different in different directions.

Suppose we were to isolate each star, and look at the shape its orbit traces out as it makes one loop around the nucleus. Some orbits would be nearly circular, staying roughly at the same radius, while others would be cigar-shaped, swooping between large and small distances during each circuit. If we consider the *mix* of orbits (how many stars are on near-circular orbits, how many cigar-shaped, etc.), it is possible for one kind of orbit to be favored over another; moreover, the mix of orbits may be different in different parts of the cluster. To interpret the velocity data, we have to "fill in" the missing part (the part parallel to the plane of the sky) using theoretical arguments. Unfortunately, if we make the wrong assumption about the true mix of orbits, we can be led seriously astray. Suppose, for example, that most of the orbits were cigar-shaped. If we looked at the edge of the cluster, we would see stars whose motion is mainly in the plane of the sky and would therefore measure a small Doppler width. However, if we looked at the center of the cluster, the stellar motion would be mainly along our line of sight, and we'd measure a large Doppler width. If we didn't correct for the unusual orbit mix, we would conclude that the velocity goes up steeply toward the center and therefore infer the presence of a massive black hole – even if none were present.

By now, most of these concerns have been laid to rest. Observations with the *Hubble Space Telescope* have vindicated Sargent and Young's original claim by producing sharper pictures of the cusp, and by revealing a disk of glowing gas orbiting the center at a distance of about 60 light-years. Spectra of this gas tell us its orbital speed; it is then straightforward to calculate that there must be a mass of about 2 to 3 billion suns in the center to hold the gas in orbit.

M87 is not one of our very closest neighbors: it lies in the Virgo Cluster of galaxies, about 50 million light-years away. Its black hole is readily detectable only because it has such a huge mass. At such long range, even a hole of 100 million solar masses would not cause a peak in the starlight discernable to optical astronomers, nor

The *Hubble Space Telescope* has revealed a disk of glowing gas orbiting the nucleus of M87 at a distance of 60 light-years. A systematic pattern of redshifts and blueshifts indicates rotation at speeds of roughly 750 kilometers per second. The central mass required to keep gas in orbit at this speed and distance is 2.4 billion solar masses. Peter Young and collaborators derived a similar black hole mass using different observational techniques in 1978.

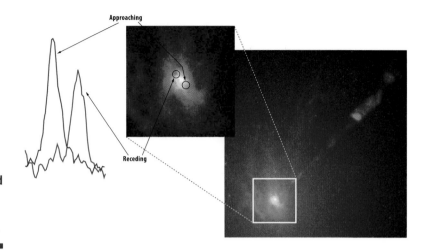

a visible quickening of stellar velocities. For this reason, astronomers searching for quiescent black holes have focused most of their attention on the nuclei of our closest extragalactic neighbors. Unlike M87, these galaxies do not currently seem "active" even at low levels, yet because they are more than 10 times closer than M87, a central mass concentration of "only" a few million suns might reveal itself gravitationally.

The most interesting and convincing case lies at the center of our nearest large neighbor in space, the Andromeda Galaxy, M31, which is about 20 times closer than M87. M31 also provides a cautionary tale about the subtleties and potential pitfalls of analyzing stellar motions in galactic nuclei. In the 1960s, Martin Schwarzschild (Karl Schwarzschild's son) and his Princeton colleagues had flown a small telescope, *Stratoscope,* in a high-altitude balloon to avoid the blurring effect of the lower atmosphere. They discovered that the stars in the central few light-years of M31's core had a flattened distribution, but they didn't guess that the explanation involved a supermassive black hole. Later measurements, by Alan Dressier of the Carnegie Observatories, John Kormendy of the University of Hawaii, and Douglas Richstone of the University of Michigan showed that the random velocities rise toward the center, and revealed an additional bonus: the entire cluster of stars is rotating. This rotation presumably accounts for the flattening of the distribution. It seemed that the detection of rotational motion would help to remove the ambiguity about the mix of stellar orbits, which bedeviled the early work on M87. Moreover, from the rotation speeds of these stars at various radii one could tell that the central mass is highly concentrated and falls in the range of 30 to 70 million solar masses.

The central region of the Andromeda Galaxy (M31) as observed by the *Hubble Space Telescope* in visible light. The stars form a highly eccentric disk around a black hole with a mass of about 140 million suns. The black hole is located close to the fainter of the two peaks in the light distribution; this location corresponds to the galaxy's center of mass.

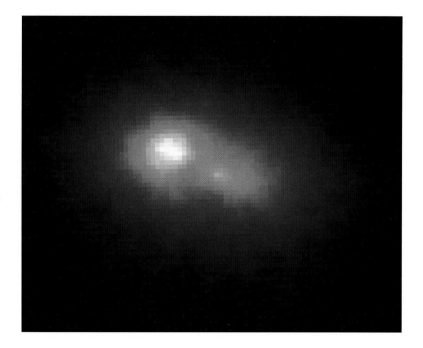

But there was a catch. Pictures of M31 taken by the *Hubble Space Telescope* revealed more detail, but in contrast to the pictures it produced of M87 the crisper images of M31 were perplexing. They showed two separate "peaks" in the light distribution, with the brighter "nucleus" offset from the true center of the galaxy. The obvious possibility that there are two black holes seemed unlikely, because two neighboring objects of such high mass would spiral together so quickly that it would be a fluke if we happened to catch the nucleus in such an unusual short-lived state. Scott Tremaine proposed an alternative explanation: that the disk of stars is highly eccentric. The brighter peak would correspond to those stars swinging through the outer parts of their orbits – because of their slower speeds they would be bunched together and therefore emit more light. Ralf Bender and collaborators confirmed this picture in 2005 by discovering that the black hole sits squarely at the galaxy's center of mass. With the new evidence, the case for a black hole is much firmer: M31 should now be regarded as the third most compelling case, after the Milky Way and NGC 4258. The estimated mass is also more reliable, and has increased to 140 million solar masses.

M31 is one of several dozen nearby galaxies for which evidence of a central black hole exists. This is a large enough detection rate to imply that massive black holes are ubiquitous in the nuclei of galaxies larger than about a tenth the mass of the Milky Way. John Tonry

This graph, based on data from the Space Telescope Imaging Spectrograph, shows orbital speeds of gas near the center of the galaxy M84. Negative velocities denote blueshifts, i.e., motion toward us, while positive velocities denote redshifts. The "zigzag" symmetry of the curve between redshifts and blueshifts is a signature of rotation. The sharp increase in orbital speed toward the center indicates a central concentration of matter weighing about 300 million solar masses.

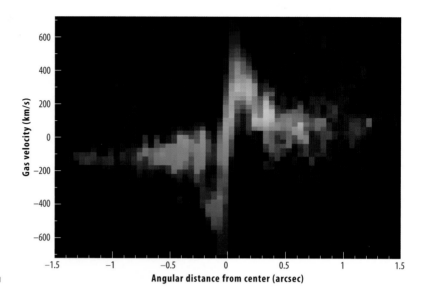

showed in 1984 that the small elliptical galaxy M32 (a satellite of M31) harbors a central dark object of about 3 million solar masses, and several other observers have subsequently confirmed his observation. As in M87, a steep cusp of stars, with random velocities increasing toward the center, reveals the presence of the compact mass. The edge-on disk galaxy NGC 3115 has a nucleus in which the innermost stars are moving unusually fast, implying a central black hole of a billion solar masses. Another interesting case is the Sombrero Galaxy (NGC 4594), studied by B. Jarvis and P. Dubath at the European Southern Observatory in Chile, and by John Kormendy with the Canada–France–Hawaii Telescope (CFHT) in Hawaii. Like Andromeda, this galaxy has an apparently rotating stellar core, but the central mass is about 7 times larger. The galaxy M84, which, like M87, lies in the Virgo Cluster, has a central disk of gas whose spectrum shows the clear "zigzag" pattern characteristic of rotation around a compact dark mass. In this case the estimated mass is 300 million solar masses. All of these results have been confirmed by exploiting the superior resolution of the *Hubble Space Telescope*.

By the late 1990s there were enough mass measurements to begin assembling a demographic survey of black holes, plotting their masses against properties of the host galaxy. Astronomers noticed a rough proportionality between the mass of the black hole and the mass of the galaxy. The proportionality is tighter if, for disk galaxies, one considers only the mass of stars in the bulge – the ratio of bulge-to-black hole mass is about 1000-to-1. An even tighter correlation seems to couple the black hole's mass to the "velocity dispersion" in the galaxy's core, a measure of the typical stellar velocity. Given the rough proportionality with mass,

The mass of the central black hole in a galactic nucleus is strongly correlated with the typical random speed of stars in the galaxy's bulge, σ ("sigma"), with mass increasing roughly as the fourth power of σ. There is also a correlation between black hole mass and bulge mass, although some studies suggest that the correlation between black hole mass and σ is tighter (and therefore, perhaps, more fundamental).

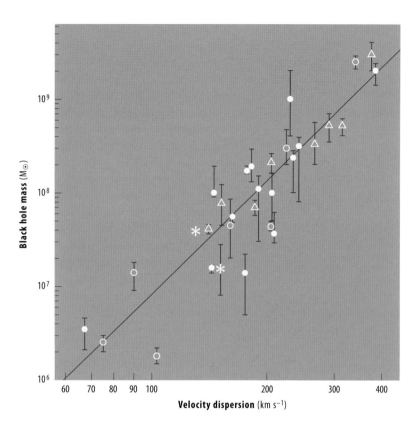

the latter correlation is not surprising, since Sandra Faber and Robert Jackson had already shown (in 1976) that the mass of an elliptical galaxy (and by extension, any bulge component of a galaxy) is approximately proportional to the fourth power of its velocity dispersion. There is still some disagreement over whether the "M-sigma" correlation (where the lower-case Greek letter σ is often used to denote velocity dispersion) is really tighter than the correlation with bulge mass, but either correlation shows a deep connection between the growth of supermassive black holes and the formation and evolution of galaxies. We will consider the nature of this connection in Chapter 9.

Are Black Holes Fussy Eaters?

It is no surprise that the "quiet" black holes in typical galactic nuclei should be on restricted diets. Their environments are not nearly as gas-rich as those found in Seyfert galaxies or quasars. A gas-poor environment should translate into a small accretion rate onto the black hole. As we saw in our discussion of X-ray binaries in Chapter 3, a lower accretion rate means that less gravitational energy is liberated and the expected luminosity is smaller.

Yet gas-poor does not mean gas-free, so even the quietest supermassive black holes should produce *some* luminosity from swallowing gas. In the center of the Milky Way we see dense molecular clouds within a few light-years of the black hole, and from X-ray observations we infer the presence of a hot gaseous medium pervading the space between them. The young, massive stars in the Galactic Center emit powerful winds, which crash into each other and may be the source of the hot gas. Once we estimate the density, temperature and velocity of this gas, we can apply the Bondi–Hoyle–Lyttleton theory of accretion – also introduced in Chapter 3 – to estimate the accretion rate and predict the luminosity.

The expected accretion rate is indeed very small, only about 1/100 000 of a solar mass per year – less than one-thousandth the accretion rate needed to power the nucleus of a minor Seyfert galaxy. The predicted luminosity is correspondingly small by the standards of AGN. But the *observed* luminosity of Sgr A* is smaller still – only 1/100 000 of the predicted value. What's going on?

Could we be seeing a break in the chain of events that converts gravitational energy to radiation as gas approaches the black hole? In accretion, the ultimate source of energy is the potential energy of the gas in the gravitational field. As gas drifts closer to the black hole, this energy is converted into a combination of kinetic energy – the energy of motion – and heat. The luminosity is produced only in the final stage as heat – the random motions of particles making up the gas – is converted into radiation. This last stage is very sensitive to density, and could be thwarted at low accretion rates. Instead of radiating away most of its thermal energy to form a thin accretion disk (as in the case of X-ray binaries, discussed in Chapter 3), the gas retains its heat as it approaches the black hole and ultimately crosses the horizon. According to this hypothesis, the accretion rate is close to the predicted value but little radiation comes out: most of the liberated energy simply disappears into the black hole.

This idea has been tested via computer simulations and detailed analysis of the radiation observed from Sgr A* and other quiet black holes. It does not hold up very well, with good reason – there are additional physical effects that tend to prevent all but a small fraction of the superheated gas from even reaching the vicinity of the black hole. Despite little net rotation, the gas in the Galactic Center probably has far too much angular momentum to fall directly into the black hole. As in a thin accretion disk, some "viscous" process must transport angular momentum outward so that the gas can drift inward. But any process that transports angular momentum must also transport energy – the viscosity acts figuratively like a "conveyor belt" carrying energy outward. But this outward flow of energy does not simply pass through the

Simulations of gas flow around black holes, such as this one by Jonathan McKinney and Charles Gammie, suggest that strong outflows are inevitable. In this snapshot, a hot torus of magnetized gas, seen in cross section, rotates around a spinning black hole, with the highest densities (red) near the equator and a nearly evacuated funnel (dark blue) along the rotational axis. Along the funnel, gas is escaping in a powerful wind.

accreting gas, unnoticed. Some of the energy liberated by gas close to the black hole accumulates in the more distant gas, and if radiation does not carry away at least two-thirds of this deposited energy, the gas grows so hot that it can escape the gravitational pull of the black hole altogether. The net result is that the black hole swallows gas at a much lower rate than predicted, given conditions in the hole's environment.

Thus, it appears that black holes are "fussy eaters." When the rate of mass supply falls below a certain threshold, only a small fraction of the supplied mass ever makes it to the hole. The low rate of mass *acceptance* exacerbates the low efficiency of producing radiation, explaining the extraordinarily low luminosities of hibernating black holes.

Flares from Tidally Shredded Stars?

We have seen how the low mass supply rates in nearby galactic nuclei can lead to even lower accretion rates and luminosities that are barely detectable, if at all. Can massive black holes lurk at the centers of these galaxies indefinitely without revealing themselves through some detectable form of violent activity, however transient? It is, of course, possible that much denser gas envelops the black hole from time to time, leading to temporary accretion at a high enough rate to produce a relatively high luminosity. One of the molecular clouds drifting through the region surrounding Sgr A* could trigger an episode of rapid accretion sometime in the next few thousand years, and there have been suggestions that the X-ray "light echo" of a luminous event several hundred years ago has been detected.

We do not know how much dense gas there is near the black hole in the Andromeda Galaxy, or in the nuclei of other nearby galaxies showing evidence for a black hole. There is no reason why these regions might not be "swept clean" of gas. The star density, however, is much better known – after all, if the stars weren't closely packed near the center of the galaxy we wouldn't have evidence for the black hole at all. The stars trace out complicated orbits under the combined influence of all the other stars and of the hole itself. But these orbits are not "fixed," like the orbits of the planets in the Solar System. They gradually change, shifted by the cumulative effect of encounters with other stars. From time to time, these encounters would shift a star onto an orbit that brings it very close to the presumed black hole; the occasional star might even plunge right in.

A star approaching a black hole faces a risk even if it misses falling through the horizon. Since the gravitational force varies with distance from the hole, there is a difference between the force acting on the side of a star facing toward the hole and the side of the star facing away. From the star's point of view there is a "tidal force" – just like the one exerted by the Sun and Moon to cause the tides on Earth – that stretches the star along the imaginary line running from the star to the black hole. If the tidal force becomes too strong, it can overwhelm the gravitational force that holds the star together – and tear the star apart. The tidal forces near the horizons of the biggest black holes are gentler than the corresponding forces near small black holes. A star like the Sun could be swallowed by the black hole in M87 while still in one piece. However, if the hole's mass were less than 100 million suns, then a star like the Sun would be tidally disrupted if it came within 10 times the hole's radius. Whether a star is susceptible to this kind of destruction depends on its density. A very compact star, such as a white dwarf, could fall inside any supermassive black hole more or less intact.

Tidal forces are bound to destroy some ordinary stars from time to time in any galactic nucleus containing a massive black hole. Every few thousand years, one of the stars would pass so close to the hole that it would be torn apart. The exact rate of these events would depend on the statistics of the stellar orbits, and particularly on how quickly stars are shifted onto orbits that take them close in. A disrupted star would send off flares that could be the clearest signal of a black hole's presence. But what would we observe when such an event occurs?

When the tidal force starts to compete with a star's own gravity, the material of the star responds in a complicated way. Because the star is moving, the direction and strength of the tidal force change rapidly with time. The star is consequently stretched along certain directions, squeezed along others, and strongly shocked. This phenomenon poses a grand challenge to experts in computer simulation. Within a few years, we should have computer-generated "movies" showing us what happens when stars of different types are tidally disrupted, and what a distant observer might measure as the observational signatures of such events. Relatively crude calculations have already been done, and they are good enough to convey the essence of what goes on.

The debris would spread out onto a range of orbits. About half would remain behind, bound to the black hole, while the rest would fan out, at speeds of up to 10 000 kilometers per second, moving on trajectories that would take it away from the hole. The most conspicuous effects would result from the bound debris, which returns to the hole and eventually swirls down into it. The nucleus would flare up to a brightness much greater than that of a supernova – almost to that of a quasar – but only for about a year. It is hard to calculate how much of this radiation would emerge in the visible, ultraviolet, and X-ray portions of the electromagnetic spectrum: candidate events have been reported in each of these bands. Another important question is: How rapidly does the brightness fade? In other words, how long does it take

A star approaching very close to a massive black hole can be torn apart by tidal forces. The disruption process violently distorts the star, removing orbital energy from some parts and adding it to others. The debris are stretched as they move out. Nearly half the debris will escape; the remainder will move on elliptical orbits bound to the black hole. The bound debris returns to the hole after (typically) a few months, at which point it can be accreted. During this transient accretion phase, a massive hole in a quiescent galaxy could become briefly as bright as a quasar. Here the size of the star is exaggerated relative to the size of the hole and the orbit.

for the last dregs of debris to be digested? All of these details are still controversial – gas flow tends to be more difficult to calculate than the movements of stars, which can often be treated as points of mass responding solely to gravitational forces.

The tidal disruption of a star is so rare in any given galaxy that we are most unlikely to witness such a "flare" from Andromeda during our lifetimes (still less from our own Galactic Center). But if most galaxies harbor black holes and we look at the nearest 10 000 galaxies, we would expect to catch a few at the peak of a flare, and probably rather more in a state where the effects of the most recent tidal disruption are still discernable. As mentioned above, a few candidate flares, mainly at ultraviolet and X-ray wavelengths, have been reported as possible tidal disruption events. A systematic search for such events would be even more valuable. A highly worthwhile program would be to monitor all galaxies in the Virgo Cluster once a month, looking for the occasional stellar disruption; another option would be to study a similar number of fainter and more distant galaxies in a small patch of sky. It would be surprising – indeed, baffling – if a prolonged search failed to discover any events.

The predicted flares may be the most robust way to confirm the ubiquity of massive black holes in nearby galactic nuclei. It takes a very powerful tidal force to shred a star. If theorists are able to predict the observable signatures in sufficient detail that we can identify the events unambiguously, the disruption of stars could offer a probe of the region close to the black hole where gravity is strong. Such probes are few and far between. Most of the evidence presented in this chapter was gathered from stars and gas orbiting far away from the central mass, where gravity is still relatively weak and speeds are much less than the speed of light. This evidence is important because it implies huge dark masses in the centers of galaxies that are too concentrated to be, for instance, clusters of faint stars. But with a few exceptions (such as the center of the Milky Way), this evidence does not allow us to rule out more extreme possibilities, such as a massive cluster composed of smaller black holes, a cluster of other stellar remnants like neutron stars and white dwarfs, or exotic hypothetical objects such as a gravitationally bound sphere of neutrinos.

Observing the tidal disruption of stars is also important because it brings us close to the strong-gravity regime where black holes display their unique properties. Einstein's theory describes – exactly – the astonishing and non-intuitive behavior of space and time close to a black hole. If we could only see this behavior directly, we could prove positively that these dark masses really are the black hole "solutions" of Einstein's equations. Are there any other diagnostics that can tell us about the region very close in? Can observations of galactic nuclei thereby enable us to check Einstein's theory? We return to this theme in Chapter 10.

9 COSMIC FEEDBACK

When the Universe was only 10 to 20 percent of its current age, or about 2 billion years old, quasars were far commoner than they are now. The bright quasar nearest to us, 3C 273, is more than a billion light-years away. Although it is hundreds of times more luminous than a galaxy, its remoteness renders it a thousand times too faint to be seen with the naked eye. In contrast, a hypothetical astronomer living during the "quasar era" would find the nearest quasar "only" about 25 million light-years away, and it would be just bright enough to be seen with the naked eye on a dark night (i.e., comparable in brightness to a 5th-magnitude star). We might have expected quasars to be closer at that time because the entire Universe had not expanded as much since its origin in the big bang. But the effect is much larger than can be accounted for in this way: quasars were a thousand times more common then, relative to galaxies, than they are today.

It has been recognized since the 1960s that every quasar must leave behind a massive black hole remnant. But are the large black holes at the centers of nearby galaxies the remnants of quasars? This question can be answered observationally.

The Rise and Fall of Quasars

It seems unbelievable that we can tell what the Universe was like billions of years ago, but there's nothing very mysterious about this. When we observe a galaxy or quasar at high redshift, we are not only seeing out to a very large distance – we are also looking backward in time. The light reaching the Keck Telescopes on the summit of Mauna Kea in Hawaii tonight might have left a quasar when the Universe was a small fraction of its present age. So, to find out what the Universe was like when it was much younger, we need "merely" to catalog the mix of objects found at distant redshifts. Carrying out such a program, however, is hard, painstaking work. Oftentimes, the objects being surveyed are close to the limits of detectability, and there is always the problem of "incompleteness": the likelihood that some objects that should be in the catalog will be misidentified or otherwise missed. In the early days of quasar research, the samples were not systematically obtained, and it was a time-consuming procedure to measure redshifts. But by now, tens of thousands of quasars have been cataloged, their luminosities and redshifts recorded.

What we find is that quasar activity peaked sharply about 2 or 3 billion years after the big bang. At that time galaxies were three times closer together, on average, than they are today. But quasars were 30 times closer together. Does this trend continue at earlier epochs? Apparently not. Astronomers have discovered relatively few quasars at redshifts much larger than the redshift at the quasar peak. Of course,

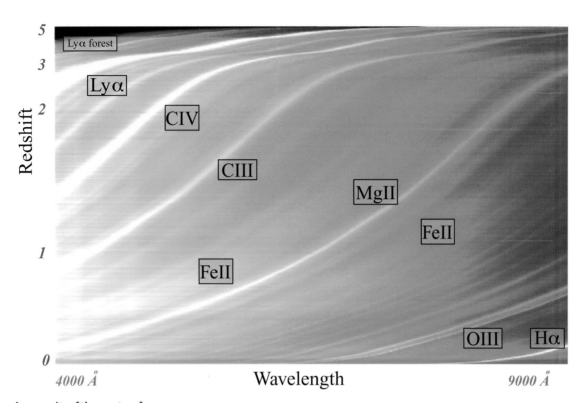

A composite of the spectra of 46 420 quasars, measured by the Sloan Digital Sky Survey. Redshifts of the quasars range from much less than 1 to greater than 5. Emission lines of various ions appear as bright bands, which migrate from short wavelength to long wavelength as the redshift increases.

very distant quasars are fainter and more difficult to detect. But this only partly explains the paucity of quasars observed in the very distant past: at larger redshifts – and thus earlier times – the quasar population genuinely thins out. Presumably, we are seeing the Universe as it was before most quasars had formed. A few quasars have been found with such high redshifts that their light was emitted when galaxies were seven times closer together than they are now. This means that the wavelength of the light has been "stretched" by a factor of 7 between emission and reception: the visible light reaching us was emitted in the far ultraviolet part of the spectrum. When we observe the remotest quasars, we are "seeing" the Universe as it was less than one billion years after the big bang. It seems, though, that most quasars appeared on the scene only after the Universe had been expanding for 2 or 3 billion years. They were relatively common for a similar period of time, and then, like the dinosaurs, they more or less became extinct.

The "luminosity function" of quasars – the relative numbers of quasars with different luminosities – yields more subtle information about their evolution. It seems that those few quasars that did form at high redshifts were predominantly the more luminous ones. The dominant quasars at lower redshifts tended to be less luminous. Recent

The quasar PC 1247+3406 has a very high redshift of $z = 4.897$. The ultraviolet lines of hydrogen (Lyman α, emitted at 1215 Å) and triply ionized carbon (C IV, emitted at 1550 Å) are observed in the red part of the spectrum. Light recorded in this spectrum left the quasar when the Universe had expanded only to one-sixth of its present scale, about one billion years after the big bang.

This graph shows how the abundance of quasars has changed with redshift. The vertical scale shows the density of quasars in space relative to their local ($z = 0$) density. Changes in density resulting simply from cosmic expansion have been removed before plotting the diagram. The pronounced peak that remains shows that quasars were most common at a redshift of 2 to 3.

observations suggest that this trend toward lower luminosity AGN – whimsically called "downsizing" – persists all the way to the local Universe where we observe the low-luminosity siblings of quasars, the Seyfert galaxies.

What does this pronounced evolution in the quasar population signify? Quasar activity presumably can't start until at least some galaxies have condensed from the expanding Universe and developed dense nuclei where matter can be fed into a black hole. During the course of a galaxy's evolution, its gas gradually becomes "tied up" in stars; quasar activity might decline because less gaseous fuel is available as galaxies age. The quasar activity might also quench itself by ejecting gas from the galaxy. Intense activity could then restart only if the nucleus suddenly acquired a fresh supply of fuel – for example, through a merger with a neighboring gas-rich galaxy.

But fuel supply is only half the story. The level of activity also depends on the mass of the black hole. We do not know whether massive black holes form concurrently with galaxies, or predate them. Are they the "seeds" around which galaxies condense? How much does the black hole grow during the period of quasar activity?

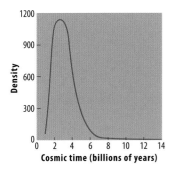

The same data as the previous figure, except that the redshift scale of the horizontal axis has been converted to "time since big bang," assuming a standard cosmological theory; also the vertical scale is linear, rather than logarithmic, to emphasize how drastic the evolution really is. The peak of the curve represents about 2 percent of the present abundance of luminous galaxies.

We can learn a lot by studying quasar statistics. Quasars are rare cosmic beasts. There are now fewer than one for every 100 000 galaxies. Even during the era of peak quasar activity, when the Universe was only 2 billion years old, quasars were 100 times rarer than normal galaxies. One might infer from this that only one galaxy in 100 has ever indulged in quasar activity, and that we shouldn't expect to find remnants in more than about 1 percent of galaxies today. But this argument assumes that individual quasars live for more than a billion years. There is another possibility. A graph like the one shown at left plots the rise and decline of the *population* of quasars: it need not represent the life cycle of a typical object. Just as in the biological case, many short-lived generations could be born, evolve, and die over the period during which the overall population rises and falls. If there were many generations of quasars, each living much less than a billion years, then we would expect to find more remnants.

The longer a quasar lives, the more massive its black hole has to be. The most powerful quasars radiate energy equivalent to several times the Sun's mass each year. The energy is released when gas swirls downward toward the hole, where gravity converts, at most, about 10 to 20 percent of the mass-energy (mc^2) of the infalling matter into quasar radiation. Thus, the hole would need to swallow at least 5 to 10 solar masses for each solar mass worth of energy that is radiated. Obviously, if a very powerful quasar kept shining for more than a billion years, it would end up weighing tens of billions of solar masses. Given the same lifetime, 90 percent of all quasars would weigh more than one billion solar masses. These mass estimates are lower limits, since they do not include the mass that might have accumulated in the black hole before the quasar epoch began. Since the lower limit on the mass is proportional to the total energy expended over the active lifetime, the final masses needn't be so high if there have been many successive generations of short-lived quasars.

Did one generation of quasars live for a billion years, leaving behind rare, enormous holes? Or did many generations flare up and fade during the quasar era, leaving behind less massive holes in most, or even all, galaxies? To answer these questions, we need an independent way to measure the masses of the black holes in quasars. Direct mass estimates are difficult to obtain, but we can bracket the most probable masses between upper and lower limits. We have reason to believe that typical quasars weigh at least 100 million suns, irrespective of their lifetime. This argument depends on the fact that the radiation produced by the quasar exerts a repulsive force on infalling gas that tends to cancel out gravity. The pressure of the quasar's intense radiation would expel everything from a quasar's vicinity unless gravity provided a strong enough counter-vailing force. The maximum luminosity for a given mass is the Eddington

This graph shows how the "luminosity function" of quasars changes with redshift. On the axes of this log–log plot, each tick-mark corresponds to a factor of 10 in either the luminosity or the density of quasars in space. Data points, from the Sloan Digital Sky Survey, represent the density of quasars per decade of luminosity for a specified range of redshifts, z. The plot clearly shows that quasars of a given luminosity become more common with increasing redshift, up to a redshift of about 3, but decline thereafter. The fact that the curves are flatter at high redshifts implies that very luminous quasars were relatively more common at earlier times, compared to less luminous ones.

limit, which we introduced in Chapter 5. If quasars are powered by accretion, and pull in gas from their surroundings, then they must have a big enough mass for gravity to overwhelm radiation pressure. Knowing the quasar luminosity, we can calculate a lower limit to the mass. By this reasoning, the Soviet theorists Yakov Zel'dovich and Igor Novikov deduced, as early as 1964 (only one year after quasars were discovered), that typical quasars had to weigh at least 10^8 solar masses, and that the most powerful ones had to weigh 10^9 solar masses or more. Another argument yields a lower limit for the minority of quasars that are radio sources: the magnetic fields and fast electrons in the radio lobes contain a huge amount of energy – the mass-equivalent of millions of suns. Whatever "engine" generated these lobes must therefore weigh at least 10^8 solar masses.

One can deduce an upper limit to the black hole mass by studying the emission from fast-moving clouds of glowing gas, which are heated by ultraviolet radiation from the vicinity of the black hole. We estimate the distances of the clouds from the black hole by measuring how much the ultraviolet light disturbs the atoms (technically, we use the spectral emission lines to gauge their level of "ionization," or how many electrons have been stripped off the atoms). The clouds around the most luminous quasars are about a light-year away from the hole; in less luminous sources, they lie at smaller distances. Using the Doppler effect, we derive the clouds' velocities from the spread in wavelengths of the spectral lines. These velocities are typically 1 to 10 percent of the speed of light. Given the clouds' distances, we can calculate how massive the black hole would have to be to generate their velocities gravitationally. Actually, we do not know whether the cloud motions are gravitationally induced, but this does not affect our ability to obtain an upper

The rise and fall of the entire quasar population is not necessarily synonymous with the life cycle of an individual quasar. The curve plotted on page 207, expanded here, depicts the number of quasars existing at different times, but these need not be the same objects. Individual quasars might live for much less than a billion years. If many short-lived generations of quasars emerge and die during the time over which the population rises and falls, then a large fraction of present-day galaxies could have been quasars in the past.

mass limit. The velocities could be larger than the gravitational value if some other force were affecting the clouds' motions, such as radiation pressure; but they are unlikely to be smaller. One therefore gets an upper limit on the central mass.

A technique called "reverberation mapping" – which relies on the fact that the ultraviolet radiation from an AGN is unsteady – has helped to firm up these results. By measuring the time delay between a flare in the ultraviolet radiation and the response of a particular emission line, one can obtain an independent (and more direct) measurement of the clouds' distance. Application of this method to Seyfert galaxies – where the delay time is relatively short – shows that different emission lines are produced at different distances from the black hole. The clouds' velocities decrease with distance roughly as expected if they were induced by gravity. It is harder to test for gravitational motions in quasars because the time delays are so long, but if we extrapolate the results from Seyfert galaxies then our mass upper limits are actually mass estimates.

We would be in trouble indeed if our derived upper limits were smaller than our lower limits! But fortunately they are not, and we can conclude that the black holes in most quasars probably have masses between about 100 million and a few billion suns. What does this tell us about quasar lifetimes? That quasar activity in any individual galaxy probably lasts for a total of tens to hundreds of millions of years, not billions of years. Although quasars never existed in more than about 1 percent of galaxies *at any given time*, we deduce that more than 10 percent of galaxies, and perhaps the majority, went through a quasar phase at some point in their evolution.

If there are enough quasar remnants to populate the nuclei of most large galaxies, do their masses agree with the results of the black hole census? We saw in Chapter 8 that the masses of black holes in present-day galaxies appear to be correlated with the galaxies' masses and/or their velocity dispersions. By measuring the distribution of galaxy properties in the nearby Universe, we can infer the "mass function" of black holes.

The luminosity function of quasars and the mass function of the black holes they leave behind are linked, because it requires accretion to produce the luminosity. Since no more than 10 to 20 percent of the energy-equivalent mc^2 is liberated when a black hole grows by m, we can estimate the *minimum* amount of black-hole mass accumulated during the quasar era by adding up the energy emitted by all quasars over cosmic history. We can then compare this lower limit to the total mass of black holes in present-day galaxies.

This conceptually simple yet profound calculation was first carried out by the Polish astrophysicist Andrzej Sołtan in 1982. Its most

The emission lines of an AGN rise and fall in response to changes in the ultraviolet luminosity, with a time delay that depends on the distance between the line-emitting gas and the black hole. By measuring the time delays for different lines – a technique called reverberation mapping – one can build up a picture of the black hole's environment. With additional information, such as the velocity of the line-emitting gas, one can measure the black hole's mass.

recent incarnations suggest that luminous accretion in quasars accounts for most of the mass contained in present-day supermassive black holes, leaving little room for growth via mergers of small black holes or accretion obscured by dust. This conclusion is sharpened if one considers the growth of black holes with a particular mass. However, there is still enough uncertainty about the local mass function, and the maximum conversion rate of mass into luminosity, that these results should be viewed with caution. In fact, a separate set of measurements, based on X-ray observations, suggests that most black-hole growth is hidden behind thick rings of dusty gas, which block virtually all visible light but let through energetic X-rays.

Like the Anasazi tribes in the southwestern United States, whose culture flourished for hundreds of years but vanished abruptly around A.D. 1300, the quasars seem to be a cosmic "culture" from an earlier era in the history of the Universe. But the descendants of the Anasazi are still among us: they are the modern Pueblo Indians. Are the descendants of ancient quasars still active in the modern Universe? Some of them could be powerful radio galaxies. These objects are rare, being found in fewer than 1 percent of large elliptical galaxies, but they do seem to be much more common than quasars at low redshifts. They do not radiate with the intensity of quasars, but their power outputs in the form of jets can be comparable. A quasar phase of activity, long ago, might have built up the mass of the black hole, and also could have given the hole spin. Perhaps the hole then lurked quiescent, the galaxy swept clean of gas, for billions of years. It took some event, maybe an encounter with a companion galaxy, to renew the infall of gas, at too low a rate to yield a quasar but at a rate sufficient to reactivate the nucleus by introducing a magnetic field to the spinning hole. This "engaged the clutch," tapping the hole's latent spin energy and converting it into jets that plow their way out to distant reaches of intergalactic space. If this is indeed what has happened in Cygnus A and M87, then these large scale manifestations of AGN activity could offer the most direct evidence for the massive black hole remnants of quasars.

Some Seyfert galaxies, and lower-powered radio galaxies, could also be remnants of quasars. Seyfert nuclei, in particular, are found in about 1 percent of modern spirals. Their spectra are very similar to the spectra of quasars, but they are less luminous. Seyferts alone could account for every quasar remnant if the typical quasar lifetime were as long as a billion years. But the masses of their black holes, where these can be estimated, tend to be lower than those inferred to exist in the most luminous quasars. One should not be surprised at this. Even during the quasar era black hole masses would have been distributed over a range extending down to fairly low values. Black holes in the lower mass range would not have produced sources of radiation easily visible at high redshift but could have survived to produce lower-luminosity AGN today.

It is very unlikely that currently active galactic nuclei account for most of the quasar remnants. There are simply not enough of them to account for all of the quasars that once existed. And many quiet, local galaxies, such as the Sombrero Galaxy and even Andromeda, host black holes massive enough to have once shone as quasars. Seyfert galaxies are probably nothing more than ordinary, "hibernating" galaxies whose nuclei have been revitalized with a fresh supply of fuel. Indeed, the same technique that has allowed us to estimate the masses of quasars – the comparison of ultraviolet luminosities with the Doppler widths of lines – has been used to extend the black hole demographic correlations to nearby Seyfert galaxies. As far as one can tell, the relationship between black hole mass and galaxy bulge mass or velocity dispersion is indistinguishable from that found in non-active galaxies.

For the more massive galaxies, at least, this relationship was apparently established in the distant past, when the galaxies were forming. It would not be surprising if a bigger galaxy hosted a bigger black hole. But what sets the ratio of black hole mass to galaxy mass at roughly 1:1000? And is it really the size of the galaxy that limits the black hole's growth, or the other way around? Surprisingly, the evidence seems to suggest that in this case, the tail may wag the dog.

Putting the Brakes on Galaxy Formation

The hierarchical theory of galaxy formation, outlined in Chapter 4, works quite well until galaxies actually begin to form – that is, until the complex processes of gas dynamics overwhelm the simple gravitational dynamics of dark matter. While the numbers and sizes of dark matter haloes, and their tendency to cluster and merge, generally agree with predictions, things begin to go wrong as gas cools and condenses into these haloes. First, cooling tends to be a runaway process: theoretically, at least, whenever it can occur it does, in an accelerating fashion. This leads to the prediction of a galaxy "obesity" problem – there should be many more huge galaxies than are actually seen. The large galaxies should continue to absorb gas from their surroundings far longer than they actually do, and should therefore continue to produce new stars, in large numbers, late into their adolescence. Observationally, this means that very large galaxies – mainly ellipticals – should have a bluish tint, from all the young stars. In reality they are red, implying that star formation stopped early. Disk galaxies are similarly expected to be bloated, with disks extending to much larger radii than observed.

The effects of feedback on the predicted distribution of galaxy masses, compared to the observed distribution (blue data points). Without feedback, galaxy formation theory overpredicts the numbers of both large and small galaxies (red histogram, left panel). Feedback from supernova explosions and starlight suppresses the formation of small galaxies, but has little effect at large masses (middle). By including feedback from black hole growth, one can fit the observed numbers of massive galaxies as well (right).

Despite a tendency for smaller structures to be swallowed up by larger ones, galaxy formation theory also predicts that the process of dissolving the small haloes should be inefficient. Each large galaxy in the present-day Universe should be accompanied by numerous – dozens, if not hundreds – small satellite galaxies. Yet little of this "substructure" has been found: the Milky Way, for example, appears to have only a few small satellites.

Taken together, this evidence suggests that some destructive agent is missing from the hierarchical theory of galaxy formation, something capable of erasing many of the smaller objects while limiting the growth of the largest ones.

This something is thought to be "feedback" – extra energy released into the cosmic environment as galaxies form – that prevents too much gas from collecting in haloes. But what is the source of this feedback? At first, astronomers suspected that stellar explosions could do the job. The most massive stars quickly evolve into supernovae, each of which sends several solar masses of debris hurtling outward at a few thousand kilometers per second. But closer inspection revealed flaws in this idea. Each explosion also releases a mixture of heavy elements into the environment. If there were enough supernovae to sweep the gas out of massive haloes, the abundances of these heavy elements would be too high. At best, supernovae could solve the substructure problem by sweeping the gas out of small haloes. In the smallest, "pregalactic" haloes, gentle heating by ultraviolet starlight could also help by evaporating some of the gas. But the massive haloes would remain unscathed – they must be subject to feedback from a completely different source.

Many astronomers believe that it is the growth of supermassive black holes that regulates the assembly of massive galaxies. How could a black hole, with a radius that fits inside the Solar System, govern the organization of thousands of times more matter spread across tens of thousands of light-years? The energy supply is more than adequate, if we assume that each black hole of mass M liberates an energy of about $0.1 Mc^2$ in growing to its final size. By this estimate, the supermassive black holes we measure in nearby galaxies collectively released enough energy to accelerate all the gas and stars in their hosts to speeds approaching 1000 kilometers per second – more than enough to tear the galaxies apart and disperse their matter.

The problem is apparently not one of supply, but of distribution. How did a fraction of the energy get spread over such large distances, and how did it interact with the gas coming together to form the galaxy? The light from a quasar – an intense beacon of ultraviolet, optical and X-ray radiation – has no problem traversing billions of light-years. Under certain conditions X-rays could heat the interstellar gas, loosening gravity's hold on it. But under most circumstances, this kind of heating would not be strong enough, nor widespread enough, to stem the growth of the galaxy. AGN put out energy in a second form, however, that is harder to resist: the kinetic energy of jets and winds.

The powerful radio galaxies we studied in Chapter 6 are rare, even by quasar standards. But there are good reasons to suspect that black holes routinely return a fraction of the energy liberated during accretion to their environments, in the form of fast streams of escaping gas. In order to produce nothing but radiation, an accretion disk has to be amazingly efficient at converting heat or magnetic energy into light. Any substantial break in the converson chain – such as the escape of loops of magnetic field through a corona, as happens in the Sun – will lead to the ejection of gas, most likely at high speed. This is just a generalization of the principle that turns a nearly starved black hole into a "fussy eater" (Chapter 8). We are not yet able to predict theoretically how much of the liberated energy is ejected in kinetic form, but we have observations to guide us. The jets of lesser radio galaxies, like M87, are examples of the kind of feedback that could regulate galaxy formation. So are the quasar winds which, though "radio-quiet," are detectable through observations of ultraviolet or X-ray absorption.

The manner in which the kinetic energy of a wind or jet couples to the environment is not well-understood. This interaction depends not only on the characteristics of the wind or jet, but also on the detailed distribution of gas in the environment. Turbulence, sound waves, and heat conduction can all play a role in spreading out the energy. We will have more to say about this later in the chapter, when we discuss feedback in clusters of galaxies. One can make some general statements about the effects of feedback on galaxy formation, based on simplified descriptions of the interaction. A lot depends on

Relative intensity

2.72   5.44   13.60   27.20
Rest-frame energy (keV)

The observation of "broad absorption lines" in the spectra of certain quasars implies that most quasars produce fast winds that inject kinetic energy into their environments – even if there is no sign of a radio jet. So-called BAL quasars were discovered at visible and ultraviolet wavelengths; X-ray observations, such as the one shown here, suggest that these winds may be even more powerful than previously thought.

how efficiently the kinetic energy is transferred to the surrounding gas. If the interaction is efficient, then even a fairly low level of feedback can have a dramatic effect on galaxy formation. However, the impact of feedback cannot be neglected even if the interaction is rather wasteful of energy. The jet or wind still possesses momentum, which will tend to push aside the gas that is trying to assemble the galaxy. These kinds of arguments have led to a number of possible explanations for the correlation between black hole mass and mass or velocity dispersion of the host galaxy. Ultimately, one arrives at a cosmic chicken-or-egg problem for which we do not yet have an answer: did the growth of the black hole stop the galaxy from growing beyond a certain mass? Or did the growth of the galaxy eventually limit the fuel supply to the black hole, limiting its mass in the process?

Forever Blowing Bubbles

One reason our understanding of galaxy formation is so incomplete is that we lack observations of the process itself. Large galaxies, along with their black holes, formed a long time ago. We see young galaxies as faint smudges at high redshift. The next generation of telescopes may bring them into sharper focus, but even so, it will not be easy to unravel the multiple, complex processes that shape galaxies. At the same time as (we think) black hole feedback is "pushing back" on the gas trying to form a galaxy, gas clouds are condensing and cooling, stars are forming, and small satellite galaxies are spiraling in and being absorbed.

Clusters of galaxies are simpler. Most of the gas in a cluster has not formed stars, nor even been absorbed into galaxies. There is no dust to obscure our view. Moreover, clusters of galaxies are still forming – we can study their formation process at relatively close range.

Clusters also show evidence for feedback. As galaxies and groups of galaxies collide to form ever larger structures, their gaseous components get shocked to higher temperatures. The gas permeating a cluster is so hot that it is best observed with X-rays; its density is so low that little of the gas has time to cool. Theory thus predicts that the density and temperature of the gas should be tightly coupled to each other: observationally, the X-ray luminosity of a cluster – which is much more sensitive to density than to temperature – should increase as the square of the temperature.

Observations do not bear out this prediction. Small clusters and groups of galaxies, in particular, show a deficit of X-ray luminosity

X-ray brightness vs. position on the sky, for a cool-core cluster (Abell 178: left) compared to a cluster without a cool core (Coma Cluster: right). The spike in X-ray brightness shows that gas in the inner part of the cluster is radiating away its thermal energy at a rapid rate. Without compensation for this loss of energy, the gas would lose pressure support and collapse into the center.

which, paradoxically, means that the gas has been subjected to *excess* heating. The extra heat increases the pressure, which puffs up the gas and reduces its density, thereby decreasing the luminosity. It is as if a fixed amount of extra heat has been added to every gram of gas. Since the smaller clusters start out with less heat per gram, the addition of a fixed amount is most obvious for these systems.

Supermassive black holes very probably supply the feedback that limits the growth of massive galaxies; they are also the most likely source of extra heating in clusters. But where the feedback in galaxies need only be distributed over thousands of light-years, in cluster feedback the energy must spread itself across distances as large as millions of light-years. The heat contained in cluster gas is a relic of the cluster's formation, and as such provides only indirect evidence for feedback. But unlike galaxies, where we see only the consequences of feedback long after it has ended, clusters also show feedback in action.

Some of the evidence for ongoing feedback comes from observations of *cool cluster cores*. Although cooling is too slow to be important in the outer parts of clusters, the situation is touch-and-go in the inner parts, where the density is larger. In about half of all clusters, the gas close to the center is dense enough to radiate away a large fraction of its thermal energy. The evidence for this is an intense spike of X-ray radiation near the cluster center. When cool cores were discovered in the 1970s, it was assumed that loss of pressure would cause the radiating gas to sink toward the center of the cluster, unable to resist gravity. As fresh gas refilled the

As X-ray-emitting gas in a cluster of galaxies cools, it emits a range of spectral lines at different wavelengths, shown in green. Observations of cool cores in clusters show some of these lines (red), but not others, implying that the gas stops cooling below a certain temperature.

A plot of temperature vs. distance from the cluster center shows that the temperature decreases by a factor of about 3, but then levels off. Without feedback, the gas temperature would drop to much lower levels near the cluster center.

central regions it would establish a "cooling flow," with (in some cases) hundreds of solar masses drifting toward the cluster center each year.

No evidence of the collapsing gas was ever detected, and in 2000, observations made with the *XMM-Newton* and *Chandra* X-ray observatories showed why. Gas slumping toward the cluster center due to unchecked cooling would drop in temperature, and as it passed through each temperature range it would emit a unique set of spectral lines. Adding up all the contributions from gas at different temperatures, one can predict the spectrum of a cooling flow. But the observed spectrum is quite different: the lines from the hotter temperatures are present, in the predicted ratios, but those from the cooler components are missing – or at least very weak. It seems that most of the gas cools just so far and no further; indeed, X-ray maps show the temperature decreasing toward the cluster center but then leveling off after it has dropped by about a factor of 3.

To stop the gas from continuing to cool requires feedback, not merely when the cluster forms but continuously thereafter. Moreover, the feedback must be regulated so that the heating of the gas balances, on average, the loss of energy due to radiation. Once again black holes are suspected, and this time the evidence that they are implicated is much more direct.

First, there is statistical evidence. We have long known that AGN – mostly radio galaxies – are often found in the centers of large

X-ray observations reveal "holes" in the central region of the Perseus Cluster. An overlaid map of radio emission (contours) shows that the energy responsible for displacing the X-ray emitting gas is being injected by jets from NGC 1275. The bright rims around the holes are slightly cooler than surrounding gas, implying that the holes are gently inflated "bubbles" rather than violent blasts.

clusters, and particularly in clusters with cool cores. We can estimate the power outputs of these jets – albeit with a good deal of uncertainty – and compare them to the powers radiated away by the cluster gas. Averaged over all the cool core clusters, the two powers – input and output – roughly agree, confirming that black holes are at least capable of producing the feedback. But the morphological evidence is far more powerful: we see the imagery of feedback on the sky.

Consider the case of the cluster of galaxies in Perseus, which contains a cool core with a radio galaxy called NGC 1275 at its center. Straddling the active nucleus are two "holes" in the bright X-ray glow, each one spanning tens of thousands of light-years. These are locations where the X-ray emitting gas has been displaced, not by cold gas but by pressurized gas with an even higher "heat" content – we know this because the holes are filled with radio synchrotron emission, produced by extremely energetic electrons in a magnetic field. Moreover, the radio emission appears to be connected to the nucleus via jets, implying that these holes are inflated by energy liberated close to the black hole.

We have already seen, in Chapter 6, that the jets of powerful radio galaxies are very effective at displacing intergalactic gas. The

Three stages in the evolution of a radio galaxy. In the "momentum-dominated" phase (top), the cocoon is long and narrow, reflecting the directionality of the jets. Eventually the cocoon widens until it becomes nearly spherical, heralding the "energy-dominated" phase (middle). After the expansion speed slows down below the sound speed in the cluster gas (about 1000 kilometers per second), the jets inflate a series of "bubbles" that "float" outward through the cluster (bottom). For long-lived radio galaxies, this is the dominant evolutionary phase.

radio lobes and hotspots of Cygnus A, for example, are created as the jets violently push aside the surrounding gas. But the situation in the Perseus Cluster is different. The bright rims surrounding the X-ray holes are not shock waves but layers of relatively cool gas, dredged up from the cluster center. Apparently the jets in Perseus are not slamming into their surroundings with violent force, but are pushing the gas aside rather gently – NGC 1275 is "blowing bubbles."

The idea of a jet gently inflating bubbles seems at odds with our picture of a powerful radio galaxy such as Cygnus A. But it does not seem so strange once we consider the life-cycle of such a galaxy. In Chapter 6, we saw how the jets from a radio galaxy bore through intergalactic gas while leaving behind large amounts of energy that inflate a cigar-shaped cocoon. In the early stages the cocoon is long and thin, mirroring the directed impact of the jets' momentum – one might therefore call this the "momentum-dominated" phase of a radio galaxy's development. But as increasing amounts of energy accumulate in the cocoon, it begins to expand sideways as rapidly as it lengthens. Eventually the cocoon becomes nearly spherical, and plows into the intergalactic gas roughly equally in all directions. Since the direction of the jet now has little effect on the shape of the cocoon, the jets' momentum is much less important than the energy it leaves behind – we call this the "energy-dominated" phase.

During the momentum-dominated phase, the cocoon of a powerful radio galaxy typically lengthens at about a tenth the speed of light. As the cocoon becomes more spherical the expansion speed gradually decreases, and eventually it drops below the sound speed in the cluster gas, about 1000 kilometers per second (0.003 of the speed of light). Once this happens, the cocoon breaks up and the energy injected by the jets begins to spread throughout the cluster. We call this third phase the "bubble" phase, because the energy released by the disrupted cocoon is carried by low-density pockets of gas. Like bubbles in a tub of water, these tenuous parcels are buoyant and therefore "float" outward through the cluster. Apparently, we have caught NGC 1275 in the act of releasing a pair of these bubbles.

The first two phases of radio galaxy evolution are over rather quickly – they typically last no more than a few million years – but the bubble phase can go on as long as the radio galaxy continues to produce jets. A long-lived source therefore spends most of its time blowing bubbles; most of the energy returned to the cluster gas takes this form. In Perseus, there are multiple holes in the X-ray emission, some quite far from the nucleus of the central radio galaxy. Presumably, these are bubbles that were released some time ago – remarkably, they seem to have maintained their integrity.

With its hot spots and narrow lobes, Cygnus A appears to be momentum-dominated in this grayscale radio image (a). But an X-ray image (superimposed as contours in (a), and shown in false-color in (b)) shows that the cocoon is nearly spherical, suggesting that Cygnus A is making the transition to the energy-dominated phase. The hot spots are clearly visible in X-rays, and careful examination shows a pair of X-ray "holes" coincident with the radio lobes.

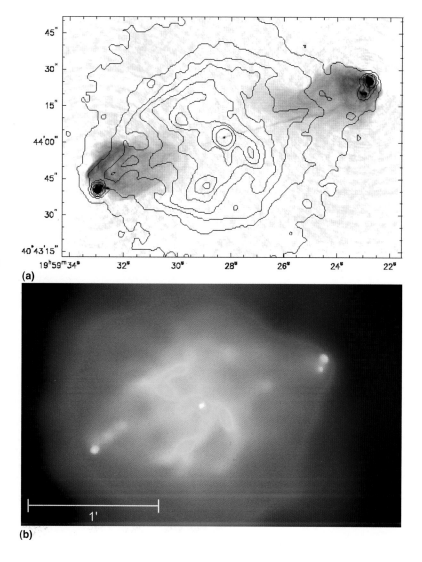

(a)

(b)

It is one thing to identify a source of energy that could lead to widespread heating of cluster gas, and quite another to understand the mechanisms by which this energy is distributed and turned into heat. This crucial stage of the energy transfer process is very poorly understood and the object of much speculation. Thermal conduction, viscous (frictional) heating, populations of ultraenergetic electrons and protons ("cosmic rays") and magnetic fields could all play a role. Much of the energy could be spread by sound waves, perhaps triggered by fluctuations in the jet power or turbulence near the walls of the bubbles. Both the Perseus Cluster and the Virgo Cluster (surrounding M 87) show concentric ripples of X-ray emission that are strongly reminiscent of such waves.

X-ray holes, seen in the outer parts of the Perseus Cluster, could be relics of earlier periods of activity in the nucleus of NGC 1275. These "fossil" bubbles could have floated away from their source under the action of buoyancy. Indications are that they are remarkably long-lived.

X-ray images of the Perseus Cluster (a) and the Virgo Cluster surrounding M 87 (b) show concentric "ripples" that are thought to represent sound waves. Ordinarily, one would expect these sound waves to amplify into shocks. Why this has not happened remains a mystery; the answer might help to explain how feedback can heat the gas so uniformly.

(a)

(b)

How is the feedback regulated, so that the energy output from the black hole neither blasts the gas out of the cluster nor allows it to cool and collapse? This, too, is a subject of great uncertainty. The elements required for a closed feedback loop are clearly present. If the gas in a cluster were allowed to cool, unchecked, large amounts of it could descend toward the center and drop into the black hole's clutches. The increased rate of accretion would then lead to an enhanced return of energy to the cluster environment. In this way, increased cooling would cause increased energy output, which compensates for the cooling and regulates it. But how this happens, in detail, is not known: the cooling properties of clusters are set tens if not hundreds of thousands of light-years away from the black hole, whereas the accretion rate is thought to be determined only a few light-years from the horizon.

Despite all these unknowns, the study of supermassive black holes in their cosmic environments has brought us to the point where we must regard these bizarre objects as key agents in the formation and evolution of galaxies and clusters of galaxies. This is a far cry from the view – prevalent until just a few years ago – that they are merely spectacular oddities in the cosmic zoo.

10 CHECKING UP ON EINSTEIN

To specialists in relativity, the black holes in binary star systems and galactic nuclei signify places where the space of our Universe has been punctured by the accumulation and collapse of large masses – collapse to smooth geometrical entities that can be described exactly by simple formulae, the "Kerr solution" to Einstein's equations. As Roger Penrose has emphasized, "It is ironic that the astrophysical object which is strangest and least familiar, the black hole, should be the one for which our theoretical picture is most complete." Black holes are, according to theory, just as "standardized" as elementary particles such as electrons and protons. All black holes of a given mass and spin are exactly alike.

It seems astonishing testimony to the power of mathematical physics that any entities of the real world can be so fully understood. This thought made a deep impression on Chandrasekhar. In a lecture given in 1975, he said: "In my entire scientific life ... the most shattering experience has been the realization that an exact solution of Einstein's equations of general relativity, discovered by the New Zealand mathematician Roy Kerr, provides the absolutely exact representation of untold numbers of massive black holes that populate the Universe." It is now even clearer, as we've seen in earlier chapters, that massive collapsed objects indeed exist, and that the power supply in quasars and other kinds of active galactic nuclei is, ultimately, gravitational. But how confident can we be that these objects have exactly the properties that Einstein's theory predicts?

It will not have escaped the reader's attention that much of the evidence for black holes is still circumstantial. That in itself does not mean that the case is weak. After all, until the detection of solar neutrinos the evidence that stars like the Sun are fueled by nuclear fusion was also circumstantial, yet we were persuaded by the success that astronomers had achieved in predicting the brightnesses, colors, radii, and other properties of stars with different masses and ages, and in explaining how nuclear reactions in stellar cores and supernova explosions transmute nuclei up the periodic table to account for the relative abundances of the chemical elements. The quest to find black holes has also had many observational successes. In X-ray binaries we measure compact masses greater than three times that of the Sun, and deduce, by a process of elimination, that these objects must be black holes. Using very-long-baseline radio interferometry, we trace the jet in M87 to within a hundred Schwarzschild radii of its presumed source. In quiet galactic nuclei – most notably the center of the Milky Way – measurements of gravitationally induced motions allow us to tighten the noose around the "massive dark objects" that have proven to be so common. The energy emerging from active galactic nuclei, and their miniature counterparts the X-ray binaries, provide strong evidence for very deep and compact "gravitational firepits," in which up to 10 percent of the rest-mass energy (mc^2) of inflowing material is converted into radiation. All of these pieces of evidence point to extremely compact masses. But they do not tell us about the detailed behavior of space and time in the domain where gravity is superstrong.

Searching for a Sign

Could our observations of quasars or X-ray binaries give us some specific sign of the curvature of space near the black hole, some signature – in the radiation emitted by gas swirling inward toward a hole, or by the jets emerging from it – that indicates whether the distortions of space and time agree with those predicted by Kerr's solution? Gas swirling into a black hole flows in a complicated way. Theorists are now computing what these flows are like, with the aim of accounting for both the radiation and the emergent jets. If the inflow is steady, the gas surrounds the hole in a configuration resembling either a thin disk or a thick donut, depending on such factors as the mass supply rate and the magnetic field. Unsteady flows – as would develop, for instance, when a star is tidally disrupted – would be much more complex, but work is underway to compute these as well. Far away from the black hole, the effects of gravity on the gas should nearly agree with Newton's theory: any deviations required by general relativity would be hard to detect. But uniquely relativistic effects should become prominent very close to the black hole. For example, according to Kerr's solution, the radius of the hole's horizon and the smallest orbit that a particle can have without falling in (which is larger than the horizon), depend on how fast the hole is spinning. The faster the spin, the closer the gas has to get before the hole swallows it, and the more energy is released (at least for gas approaching the hole in a thin disk).

If we could view the flow pattern close up, we could tell whether it is consistent with what theory predicts; if it is, we could even infer how fast the hole is spinning. A disk viewed obliquely would appear distorted by the severe bending of light. The far side would seem to be warped upward, and the disk would appear brighter on the side where the flow comes toward us than on the receding side. The hole would cast a "shadow," spanning about 5 Schwarzschild radii, against the background glow.

Such an image, viewed on the sky, would be tiny. Among known black holes, the one at the Milky Way's center would have the largest angular size, with a diameter of about 30 microarcseconds – the size of a newspaper headline seen at the distance of the Moon. The giant

In this figure, adapted from J.-P. Luminet, the right-hand image shows the appearance of a thin disk viewed at an angle of 10 degrees to its plane. The left-hand side shows the ray paths as they would appear viewed from the side. The bending of the ray paths as they pass over the hole causes an apparent warping of the far side of the disk, and the Doppler effect associated with the orbital motion causes an asymmetry between the left and right sides of the image. The bottom face of the disk's far side can also be observed (lower thick arc in the right-hand figure). The thin upper arc hugging the hole represents the bottom face of the near side of the disk, viewed via rays that are initially directed away from us but are deflected around and over the hole. (This upper arc is seen more clearly in the figure on page 227.)

black hole in M87, 1000 times more massive but 2000 times further away, would tie for second place with Andromeda (M31), while the black hole stellar remnants in our Galaxy, though relatively nearby, would appear thousands of times smaller. Thus, to image a black hole, one should focus on the Galactic Center, followed by M87 and M31.

We may not have to wait too long for the first observations. Very-long-baseline interferometry, operating at short radio wavelengths of a millimeter or less, might be able to penetrate the gas enveloping Sgr A* and thus provide the first image of a black hole's shadow within the next decade. To obtain the required angular resolution, the radio telescopes would have to be thousands of kilometers apart, but at shorter wavelengths the required separations are smaller – only tens of meters at X-ray wavelengths. With this in mind, Webster Cash, of the University of Colorado, has proposed a design for an X-ray interferometer in space, using dozens of satellites flying in formation. The proposed *MAXIM* project (for *Micro-Arcsecond X-ray Imaging Mission*) is decades away from possible realization, but might someday produce images like that shown above.

Even without super-high-resolution imaging, we can learn something about spacetime curvature just by studying the total radiation emitted from the disk. We saw in Chapter 5 how optical spectra of light from hot gas in AGN reveal Doppler shifts, which tell us how the gas is moving. The gas emitting the lines at visible wavelengths, however, lies relatively far from the black hole and moves, at most, at a few percent of the speed of light. This gas is too far away to contain a strong imprint of the black hole's geometry. The gas really close to the hole is so

An artist's conception of *MAXIM* (the *Micro-Arcsecond X-ray Imaging Mission*), a proposed X-ray telescope designed to image accretion disks around black holes.

This highly idealized computer simulation shows what a disk of gas around a nonrotating black hole would look like if it could be observed sharply enough to see the detailed structure. The disk is viewed obliquely, at an angle of 30 degrees to its plane. Because of the orbital motion, the left-hand side is blue-shifted and brightened compared to the (receding) right-hand side. In this rendering, the colors refer only to intensity, with blue being more intense, red less intense. The "arc" of light over the black hole is due to rays from the underside of the foreground disk which are deflected by almost 180 degrees and pass through the gap between the disk and the hole's horizon. The ring interior to this arc is due to photons that orbit the black hole once before escaping toward our field of view. In a more realistic simulation of an accretion disk – now becoming possible – the arc and ring would probably be obscured by the infalling gas, and turbulence would imprint a chaotic texture on the disk's appearance.

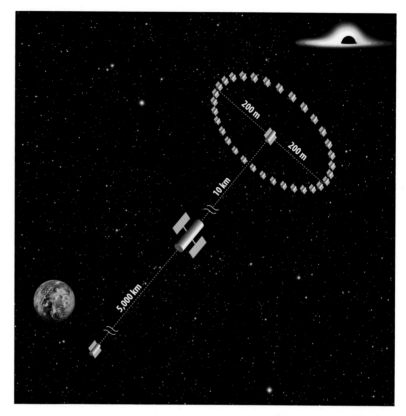

hot that it emits X-rays rather than visible light. These X-rays have been observed and are found to vary irregularly in intensity – active galactic nuclei display a scaled-up version of what happens in X-ray sources like Cygnus X-1 in our own Galaxy.

The ultrahot gas near a black hole is at temperatures of millions, or even tens of millions, of degrees. At these temperatures, much of the X-ray emission is concentrated in specific spectral lines, just like the more familiar visible and ultraviolet light from gas at 10 thousand degrees. X-ray spectrometers, such as those carried by the Japanese *ASCA* satellite (launched in 1993) and the European *XMM-Newton* satellite (launched in 1999) have measured the substantial Doppler shifts expected from gas swirling close to the hole. If the disk were viewed obliquely by one of these spectrometers, we would detect a broad line with two "wings," one blueshifted and the other redshifted, incorporating X-rays from the approaching and receding sides of the disk, respectively. The line would appear lopsided: the blueshifted side should

Most of the radiation from an accretion disk comes from its innermost regions. Photons emitted with a well-defined energy – that is, in a spectral line – will be received at a range of different energies, because the Doppler shift and gravitational redshift are different for different parts of the disk. If the disk is viewed face-on, the line is redshifted: this is partly the gravitational redshift, and partly the effect of time dilation in the rapidly orbiting material. If the disk is viewed obliquely, the Doppler effect leads to two peaks, one from the approaching and one from the receding side; the blue (approaching) peak is stronger. Because there is also a gravitational redshift, the peaks are not symmetrical about the unshifted energy; the line center is shifted toward the red. The graph at right plots the strong line due to iron in the nucleus of the Seyfert galaxy MCG−6-30-15 (adapted from A. Fabian *et al.*). Error bars indicate the uncertainty in each measurement; the red curve shows what is expected.

be much stronger because it has been amplified by the Doppler beaming effect (see Chapter 6). Finally, despite the symmetry of the disk, the wavelength of the line "center" would appear to be offset from its normal value, toward the red, even after correcting for the beaming effect and the cosmological redshift of the galaxy. This shift comes in part from the redshift that, according to general relativity, is undergone by all radiation emitted near a strongly gravitating object, and in part because fast-moving clocks (and, therefore, vibrating atoms) appear to run slow from the viewpoint of a stationary observer. X-ray astronomers have found very broad and asymmetrical lines, very similar to the theoretical predictions, from AGN as well as X-ray binaries. Although it has proven difficult to disentangle the complicated effects of accretion disk physics from the "cleaner" consequences of spacetime curvature, researchers are optimistic that these lines will provide useful diagnostics of both.

Even more information can be obtained by measuring the rapid changes of the X-ray signals with time. The *Rossi X-ray Timing Explorer*, launched at the end of 1995, carried a special purpose instrument designed to clock the arrival times of individual X-ray photons with an accuracy of several millionths of a second. Amidst the noisy flickering thought to characterize X-ray binaries, the *Rossi Explorer* revealed dozens of nearly periodic signals, called "quasi-periodic oscillations" or QPOs. Some of these oscillations remain remarkably steady, despite wild swings in the intensity or spectrum of the source. Do these QPOs carry information about the geometry of space close to a black hole or neutron star? Some theorists think so, but the nature of the connection is not yet clear. One possibility is that we are witnessing the ringing of the inner parts of accretion disks. The extreme curvature of

"Quasi-periodic oscillations" are nearly regular signals that are buried in the seemingly chaotic flickering of some X-ray binaries. For the three binaries shown here, the QPO frequencies are unchanged as the spectrum and intensity of the X-rays varies between the left and right panels. While the frequency is thought to encode information about the geometry of spacetime near the black hole or neutron star, the origin of the oscillations is unknown.

space traps small oscillations near the center of the disk, allowing them to build up until the vibration is detectable. Other possibilities include a rapid wobbling of the inner part of the disk, orbital motion of disk "hot spots," or periodic "echoes" due to flashes of light making multiple orbits around the hole – these effects would depend strongly on the black hole's spin.

Two of the effects we have already seen in active galactic nuclei could depend on distinctive properties of a spinning (Kerr-type) black hole. One is the emergence of jets from the very centers of AGN. Part of the spin energy of a black hole can be tapped to power these jets. Roger Penrose first suggested, in general terms, that such energy extraction was possible; Roger Blandford and Roman Znajek later showed explicitly how the energy stored in a spinning hole could be channeled into electrical and magnetic fields, and perhaps into a fast outflowing jet, as explained in Chapter 6. The other effect is the seemingly rock-steady orientation of jets in many radio galaxies, lasting over periods of hundreds of millions of years, enough time to build the lobes in powerful sources. Such stability would be hard to understand if there were no stable gyroscope to enforce it. But a Kerr black hole is the perfect instrument to provide just this kind of stability.

The direction in which a jet travels is thought to be set by the rotation axis of the swirling gas just outside the black hole. In Newtonian theory, this rotation axis does not change as the gas falls toward a gravitating body. Its direction is fixed with respect to what physicists call an "inertial frame" of reference, that is, one that is not rotating. When a physics demonstrator sets a toy gyroscope in its cage spinning in a classroom, the "inertial frame" is effectively defined by the classroom. No matter how the demonstrator tips the cage, the gyroscope will keep the same orientation with respect to the walls of the room. (In reality, a gyroscope maintains its orientation with respect to the "distant stars," slowly changing its spin axis as the Earth rotates. But this effect is seldom noticeable in classroom demonstrations.) In the same way, Newton would predict that the spin axis of the gas just outside a black hole would be the same as its spin axis far away from the hole, where it first starts drifting in. But if this were true, the direction of the jet would change every time the black hole accreted a gas cloud that approached the galactic nucleus from a different direction. (It is tempting to argue that the direction of the jet should coincide with the rotation axis of the entire galaxy, since the galaxy is supplying gas to the nucleus – but observations show that this is generally *not* the case.)

Einstein's theory predicts that gyroscopes behave quite differently in a gravitational field. According to general relativity, the "inertial frame" can actually be affected by gravity, especially in the strong gravitational field of a compact rotating "body" like a Kerr black hole.

In the early 1960s, Roger Penrose stimulated a "renaissance" in relativity theory by introducing new mathematical techniques that could deal with situations where symmetry was lacking and Einstein's equations couldn't be solved exactly. With Hawking, he used these methods to prove theorems about when "singularities" would occur. Penrose has contributed to many other topics in general relativity, and was the first to demonstrate that energy could be extracted from the spin of a black hole.

The effect is called the "dragging of inertial frames" or, sometimes, the Lense–Thirring effect, after its discoverers. A gyroscope, placed near a rotating body, will maintain its orientation with respect to the "inertial frame" in its vicinity, but that inertial frame will be rotating with respect to a distant observer. So, the distant observer will see the gyroscope's axis change with time! In a sense, the Lense–Thirring effect causes the *space itself* near a Kerr black hole to rotate in the same direction as the hole, with an angular velocity that increases as the hole is approached.

What does this mean for matter flowing toward a Kerr black hole? First, it means that nothing can fall straight into a Kerr black hole along a radius. If you dropped a rock (with zero angular momentum) into a spinning black hole, you would observe it to be deflected from its radial plunge before it disappeared beneath the horizon. No rocket engine, strapped to the rock, could develop enough thrust to stop the deflection once the rock crossed into the hole's "ergosphere," within which the dragging of inertial frames becomes irresistible. If the rock were in a circular orbit that was not lined up with the equatorial plane of the hole, the Lense–Thirring effect would cause the plane of the orbit to wobble, or "precess," about the hole's rotation axis. The same effect also causes the gas orbiting the hole as part of an accretion disk to precess. Since orbits at different radii precess at different rates, the Lense–Thirring effect would tend to distort a misaligned disk and induce some degree of chaos in the flow. However, the viscosity of the disk tries to smooth it out and restore its planar structure. The competition between these effects ultimately nudges the disk's axis into alignment with the rotation axis of the black hole. Thus, the hole really does act like a gyroscope, enforcing a steady orientation on any flow pattern, whether of disk or jets, near it.

Unfortunately, the behavior of the ultrahot gas and electromagnetic fields around a black hole is harder to understand than the hole

Direction of spin

Black hole

Orbits of objects near black hole

The path of an object falling toward a Kerr black hole is influenced by the rotation of the hole, shown here as counterclockwise. The yellow particle approaches along a radial trajectory, but its path is deflected into a counterclockwise spiral. The blue particle approaches on a clockwise orbit; a distant observer sees its direction reverse so that it too spirals into the hole on a counterclockwise path.

itself. Although one may suspect that Lense–Thirring precession and the Blandford–Znajek process are important in AGN, it will not be easy to bolster this belief by really compelling observational evidence. One would have thought that it would be easier to perform "clean" tests of general relativity when black holes rather than less compact objects like the Sun are involved, since the effects of relativity are large (rather than differing from Newtonian predictions by only one part in a million, as in our Solar System). But measurements in our Solar System achieve precisions exceeding one part in 10 billion – these precisions are more than ten thousand times better than is needed to detect the expected effects, and therefore enough to provide a test with a precision exceeding one part in 10 000. In contrast, there are no phenomena yet observed in AGN that can tell us the properties of the central hole with a precision of better than one part in 10. But astronomers as a group possess great reserves of optimism, instilled by repeated good luck as well as the profound developments in observational techniques over the past forty years. Below we will describe other types of phenomena, involving stars and black holes in very close orbits, that might soon allow us to test relativity in the "strong gravity" regime.

Stars in Relativistic Orbits

Slowly moving bodies orbiting at large distances from a central mass follow paths that are little different from the standard elliptical orbits deduced from Newtonian theory. But general relativity does predict one small difference – there should be a very slight precession of the orbits that is missed by Newton's theory. This orbital complication is unrelated to the precession caused by the dragging of inertial frames, and occurs even for orbits around nonrotating bodies. In our Solar System, this effect appears as the famous "anomaly" in the orbit of Mercury – the precession of its perihelion, or point of closest approach to the Sun – which was convincingly explained by Einstein's general theory of relativity.

Near a black hole, orbiting bodies move at a large fraction of the speed of light, and rather than showing a slight precession their orbits trace out complicated patterns. These patterns depend on the size and eccentricity (i.e., ellipticity) of the orbit; they also depend on how fast the hole is spinning because the Lense–Thirring effect tends to drag the orbit around in the direction of the hole's spin as though the space itself were spinning like a whirlpool. We have already seen that, in a galactic nucleus, stars must occasionally venture so close to the black hole that they are either swallowed whole or torn apart by tidal forces. The former is likely to have little observable consequence, while the

Spinning Black Holes

A spinning (Kerr) black hole is more complicated than its nonrotating (Schwarzschild) counterpart. Its properties depend on two characteristics – spin and mass. Moreover, the space around it is not spherically symmetrical: like any spinning star or planet, it has an equatorial bulge. Space is not only rushing into the hole but is also swirling around the rotation axis. Two distinct surfaces are important in describing Kerr black holes. Inside the outer surface, known as the *static limit*, space swirls around so fast that even light has to rotate with the hole. Within the inner surface light is sucked inward, just as happens inside the Schwarzschild radius of a nonrotating black hole. This inner surface is therefore the *horizon*. The region between the horizon and the static limit is called the *ergosphere* (from the Greek word for energy), so called because processes in this region can extract energy from the hole.

It is possible for an object to orbit closer to a Kerr black hole than to a Schwarzschild black hole, so accretion disks around Kerr black holes can release more energy for a given accretion rate. The efficiency of energy extraction, measured as a percentage of mass converted to energy, ranges from 6 per cent for a Schwarzschild hole right up to 42 percent for a Kerr hole with the maximum allowed amount of spin. The nonrotating Schwarzschild solution is obviously a special and atypical case. However, what is remarkable is that Kerr's equations (though they only have one extra parameter, the amount of spin) are general enough to describe all black holes. A black hole may form by a general asymmetrical

The ergosphere of a rotating black hole is the region between the horizon and the "static limit," the oblate surface that surrounds the horizon in this exploded view, touching it only at the poles of the rotation axis. In the ergosphere no photon or particle of matter can resist the rotation of the black hole.

collapse, and there will be a brief phase of violent disequilibrium, during which intense gravitational waves are generated. But when the hole has settled down, Kerr's equations describe it exactly. This is the famous theorem that "black holes have no hair." Once you know the mass and spin, all other properties are fixed.

When a black hole swallows something, it gains mass and its spin generally changes. But no other traces survive of what was swallowed – this information is lost. The area of the horizon is like "entropy" in thermodynamics and always grows. For a Schwarzschild hole this area scales with the square of the mass. There is also a limit to how much energy can be extracted by crashing black holes together – the resultant single hole must have a bigger area than the separate pre-merged holes added together.

The area of a Kerr hole depends on its spin as well as its mass. It is possible for processes in the ergosphere to slow down a hole's spin and thereby actually extract energy from it, without decreasing the area of the horizon. Many astrophysicists believe that the jets described in Chapter 6 can be energized by the extraction of energy stored in the spin of the hole.

latter is too "messy" to make a good test of relativity. Is there any chance that a star could migrate into an orbit close enough to a massive black hole that relativistic effects would become obvious, without it moving so close that tidal forces destroy it? Could we ever observe the relativistic gymnastics of such a star?

There is certainly "room" for a star to follow such an orbit, provided that the black hole is sufficiently massive. A hole weighing as much as a 100 million suns has a Schwarzschild radius twice as large

Subjected to the dragging of inertial frames near a spinning black hole, the inner parts of an accretion disk become aligned with the hole's equatorial plane. This effect was discovered by James Bardeen and Jacobus Petterson in 1975.

as the radius of the Earth's orbit round the Sun and a tidal-destruction radius (for stars like the Sun) only a few times larger. A normal star would have to stay well outside the tidal limit to avoid suffering internal distortions that could mask purely relativistic effects. Around less massive holes only white dwarfs and neutron stars could safely populate relativistic orbits. But some galactic nuclei contain even bigger black holes, with horizons as large as the entire Solar System. Since tidal forces near the horizon are weaker for larger holes, an ordinary star could orbit such a hole at a few Schwarzschild radii, sustaining little damage.

To get into one of these orbits, a star would have to lose an enormous amount of energy and orbital angular momentum. How can this happen? An individual star could undergo a series of close encounters with other stars, whereby orbital energy is transferred from one star to the other. Or there could be a collective interaction involving swarms of stars. The stars that orbit close to the black hole at the center of the Milky Way, for example, originally might have formed the tight core of a cluster of stars that plunged in from a much greater distance. Their relative youth, though – less than a few million years – means that these stars must have formed relatively close to the black hole, probably not more than a few light-years away. It is also

Six consecutive orbital loops around a rapidly spinning black hole, as calculated by Kevin Rauch. The highly elliptical orbit starts out with an orientation 45 degrees away from the equator of the hole. Each successive loop can be approximated by an ellipse, but the three-dimensional orientation of each loop is different. Two effects of general relativity cause the orbit's orientation to change. First, the dragging of inertial frames, due to the hole's rotation, causes the orbital plane to precess about the spin axis. Second, the same effect that causes the precession of Mercury's orbit causes the point of closest approach to precess around the hole. When these effects occur together, they lead to the complicated orbital trajectory shown.

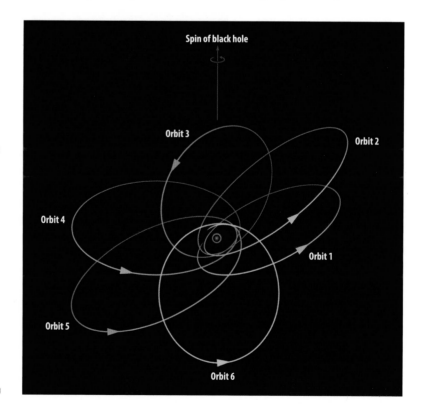

When a star crashes into the disk it punches a hole through it, and the impacted material sprays out at high speed. If the star returns repeatedly to the disk, crashing through it twice each orbit, it will gradually lose orbital energy and may eventually be captured into a very tightly bound orbit close to the black hole.

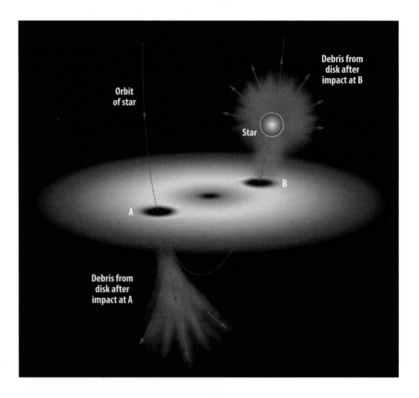

possible, then, that the stars condensed out of a gas cloud as it plummeted toward the black hole.

The presence of a dense gaseous disk orbiting the black hole – in an AGN, for example – opens up a new route for capturing stars into relativistic orbits. Twice each orbit, the star would crash through the disk. Each impact with the disk would dissipate some of the star's kinetic energy, so its orbit would gradually shrink. It is therefore possible for a star to get into the kind of orbit depicted on page 233. But would we "see" it, once it got there?

At the distance of our own Galactic Center we could, perhaps, track a star on a relativistic orbit directly. One star has already been observed to swing within 600 Scharzschild radii of the hole, reaching speeds of at least 12 000 kilometers per second (4 percent of the speed of light). The curvature of space should rotate the long axis of the orbit by a fraction of a degree every 15 years, as the star completes a circuit. This relativistic effect should be detectable with the next generation of ground-based optical/infrared telescopes, which will have apertures of 30 meters or more. Such a star would be too faint to be seen directly in other galaxies. But information about stars on relativistic orbits could still be obtained by observing the activity of AGN. A star's periodic impacts with the gas disk could modulate an AGN's power output. For instance, the debris from the impact might create an opaque screen, which would attenuate the radiation emitted by the disk for part of each orbit. Alternatively, whenever the orbiting star passed through the base of a jet, it might "short out" the electric field that accelerates the outflow. If an AGN displayed any regular periodicity (rather than just random variations), it would be hard to explain except as the work of an orbiting star. Regular monitoring of such an object should disclose not only the basic period, but also the timescale for orbital precession – thus providing a probe for the curvature of space around the hole. In the early 1990s there was a claim that the X-rays from a particular Seyfert galaxy, NGC 6814, varied with a period of 3.4 hours. Disappointingly, this period turned out to belong not to the AGN itself, but to an X-ray-emitting binary star that lay close to it in the sky. But that "false alarm" doesn't diminish our motive for continuing to search for periodicities, nor the chance that we will someday find one.

Gravitational Waves

A star could persist in a relativistic orbit around a supermassive black hole for hundreds or even thousands of years. No such orbit could persist indefinitely, however. Even if there were no drag due to interaction with gas, the orbit would gradually lose energy and the star would spiral into the hole. The reason for the star's final plunge is a fundamental prediction of Einstein's theory – the emission of *gravitational radiation.*

If the size or shape of an object changes, then so does the gravity around it. The only important exception – as Newton had realized – is that the gravitational field outside a sphere stays the same if the sphere simply expands or contracts. Changes in a gravitational field cannot propagate instantaneously – if they did, they would convey information about the size and shape of the gravitating mass at speeds faster than the speed of light, which is impossible. If the Sun were to change its shape, and thereby alter the gravitational field around it, 8 minutes would elapse before the effect was "felt" at the Earth. At very large distances, this change manifests itself as "radiation" – a wave of changing gravity, moving outward from its source. In some respects the creation of gravitational waves resembles the way a change in the electric field around a group of charges leads to electromagnetic waves. For instance, a spinning bar with a charge at each end creates an electric field that is different when the bar is end-on and sideways-on. At large distances from the bar, this changing field becomes an electromagnetic wave.

There are two important differences between electromagnetic and gravitational waves. First, any gravitational wave is very weak unless

When two black holes or neutron stars closely orbit each other, gravitational radiation will cause them to spiral together and eventually to coalesce. The waves are traveling distortions in space, illustrated here as ripples in a rubber membrane, which move outward with the speed of light.

large masses are involved. Atomic-scale "spinning bars" – diatomic molecules – are efficient emitters of electromagnetic waves (in the radio or infrared band, depending on how fast they are spinning). But they are hopelessly weak emitters of gravitational waves. As we emphasized in Chapter 1, gravity is such a weak force that even its direct effects are nearly undetectable for laboratory-size objects; *changes* in a gravitational field are far harder still to detect. Gravity becomes competitive with electromagnetism on large scales only because there are no negative "gravitational charges" (i.e., negative masses) to cancel out the positive ones, as happens, to a high degree of precision, for electricity. Large spinning dumbbells do exist in the Universe – there are huge numbers of binary stars, each containing 10^{57} atoms; and black holes, some of vastly greater mass, may in some instances have binary companions. These are the most straightforward sources of gravitational waves.

While the absence of negative "gravitational charge" gives gravity an "edge" over electromagnetism for very massive objects, it paradoxically also weakens the ability of an object to produce gravitational radiation. Here we come to the second important distinction between gravitational and electromagnetic waves. The most efficient way to produce electromagnetic radiation is for the "center of electrical charge" to wobble relative to the center of mass. This happens, for instance, when the charge on a spinning bar is positive at one end of the bar and negative at the other. Radiation produced in this way is called *dipole radiation*. But everything exerts a gravitational force exactly proportional to its mass or inertia – this is Einstein's famous "equivalence principle," according to which the gravitational acceleration, at any point, is the same for all bodies. This means that the "center of gravitational charge" is *identical* to the center of mass. The former cannot, even in principle, wobble with respect to the latter. Thus, there is no such thing as dipole gravitational radiation. Gravitational radiation resembles the case when there is a positive charge at each end of the bar: the "center of charge" remains fixed at the center, and much less radiation is then emitted because all that changes is the so-called *quadrupole moment,* which measures the extent and shape of the charge distribution. Quadrupole radiation is intrinsically weaker than dipole radiation.

As a consequence of gravitational radiation, a binary system loses energy and its orbit shrinks. This makes the stars speed up and the orbital period decrease. The effect is extremely small when the members of the binary are widely separated and orbiting each other slowly. But as the orbit shrinks, the rate of energy loss accelerates, eventually driving the binary stars toward coalescence.

The predicted rate of orbital shrinkage caused by gravitational radiation has been verified observationally, with high precision. A pulsar in a close binary system was discovered in 1974 by Russell Hulse and Joseph Taylor, using the giant Arecibo radio telescope, while they were conducting a general survey for pulsars. They immediately saw that the pulsar was peculiar in that its pulse period seemed unsteady. Hulse and Taylor quickly concluded, however, that it wasn't the pulsar itself that was unusual, but rather the pulsar's environment – it was orbiting around a close companion. Its unsteady period was a consequence of the changing Doppler effect, as the pulsar first swung around its companion toward us, and then away from us. The pulsar has an 8-hour orbit with a high eccentricity – that is, the distance from its companion changes significantly during each orbit.

For more than thirty years Taylor and a succession of collaborators have pursued a sustained program of measurements, recording the arrival times of pulses with a precision of around a millionth of a second. These cumulative observations have revealed that the orbital period is gradually decreasing, by 2.7 parts in a billion (0.27 millionths of a percent) per year. The decreasing period tells us that the orbit is shrinking, although by only a few centimeters per year. Such is the precision of the observations, however, that this tiny effect has been measured to about 0.2 percent. Moreover, it agrees, to that same precision,

The Hulse–Taylor binary pulsar follows a highly elliptical orbit around its companion star. As a result of gravitational radiation, its orbit is becoming smaller and more circular. In 300 million years, the stars will spiral together and merge. Shrinkage of the orbit can be measured with high precision by timing the closest approach of the two stars, using the pulsar as a clock. If the orbital period were fixed, the points would lie along the horizontal line in the graph. Instead, the orbital period is decreasing exactly as predicted by general relativity (red curve).

with what should happen if Einstein's theory is correct. This is in fact the only empirical corroboration that gravitational radiation is a real phenomenon. And it is the only observation that probes a distinctive prediction of Einstein's theory, one that does not appear as a correction to Newtonian theory in Solar System tests. An even tighter binary system, containing *two* pulsars, was discovered in 2003, and should provide even more precise tests within a few years.

The most dramatic bursts of gravitational radiation should occur when two compact stars in a binary – neutron stars, black holes, or one of each – coalesce. The coalescence of two neutron stars, lasting only a fraction of a second, would be a violent event that triggers a release of neutrinos (as in a supernova) and perhaps also a flash of gamma rays. If two black holes spiral together, the outcome is, in a sense, simpler than the merging of two neutron stars. The only relevant force is gravity, or "curved empty space" in the language of general relativity. No "messy" physics involving nuclear matter or magnetic fields plays any role. Everything that happens is simply a consequence of Einstein's equations. When two black holes come together, their separate horizons merge into a single, larger one, and the original binary ends up as a solitary black hole. After it has settled down, this hole must, according to the "no hair" theorems, revert to the simple, standardized form described completely by Kerr's solution of Einstein's equations. The hole ends up with a mass and a certain amount of angular momentum (or spin), but once these two numbers are fixed we can be sure that the hole is a clone of any other hole having the same mass and spin.

Frame from a simulation of gravitational waves produced by merging black holes. The curtain-like structures are the gravitational waves and the color denotes the amplitude of the waves, with yellow being the highest amplitude. The merged black hole can be seen in the center. This calculation was done by the Numerical Relativity Group at NASA's Goddard Space Flight Center using the Columbia Supercomputer at the NASA Advanced Supercomputing facility at Ames Research Center.

It is during the merger of the two holes that gravitational waves are produced in abundance. The gravity field is then fluctuating wildly; one can imagine the fabric of space itself being violently shaken about. By the emission of gravitational waves, the merging structure shakes off its "hair," the asymmetries and inhomogeneities that depend on exactly how the merger took place. It thereby erases most traces of its history, and settles toward a new equilibrium as a Kerr black hole. The gravitational waves produced in this "settling down" process vary with time in a far more complicated way than the nearly periodic waves emanating from the premerger binary. Only within the last few years have theorists – armed with teraflops of computer power and innovative algorithms – managed to solve Einstein's equations in such "dynamical" situations, when the curvature of space is changing so violently. These calculations are revealing how the detailed patterns of gravitational waves depend on such factors as the ratio of the black hole masses and the magnitudes and relative orientations of their spins. For the first time, we are able to calculate the fraction of the original mass radiated away during the settling down process, and the amplitude of the asymmetrical "kick" that can dislodge the merged black hole from the galaxy's center.

Even without detailed calculations, we can place general limits on the amount of gravitational wave energy that can be radiated during the merger of two black holes. Stephen Hawking proved, very generally, that the sum of the areas of the horizons cannot decrease. This means that when two black holes merge, the horizon around the final hole must have an area equal to or bigger than the sum of the original two. The area of a black hole's horizon depends on both mass and spin; for a nonspinning (Schwarzschild) hole, whose radius is proportional to mass, the area scales as the square of the mass. So, by placing a lower limit on the area of the final hole's horizon, Hawking also set a minimum to the final hole's mass. Suppose, for example, that two identical Schwarzschild holes, of 1 solar mass each, collided head-on to produce a bigger nonrotating hole. According to Hawking's law, the mass of the final hole must then be equal to *at least* the square root of 2 (that is, $2^{1/2}$) solar masses. Gravitational waves couldn't, therefore, carry away more than ($2 - 2^{1/2}$) solar masses, equivalent to about 29 percent of the mass of the two original holes. For a Kerr hole, with spin as well as mass, the arithmetic is slightly more complicated, but the results are similar. In practice, the amount of gravitational wave energy is often far lower than the theoretical maximum predicted by Hawking's law. This is true for the head-on collision of two Schwarzschild holes, a relatively simple problem, but also seems to hold for the more complex cases that are now being computed.

Can Gravitational Waves Be Detected Directly?

To assure ourselves that gravitational radiation is real, it would obviously be more convincing if the waves could be detected directly. Right now, the sole indication that gravitational radiation truly exists is the rather indirect evidence that orbital energy in a binary pulsar has drained away at a rate consistent with that expected. How can such waves be observed? The gravitational wave is a signal, moving at the speed of light, telling us that the shape (and hence gravitational field) of its source has changed – it would be, in effect, a slight distortion of space. The relative positions of freely moving masses, placed in the path of the wave, would jitter when the wave passed them, as though they were exposed to a varying force acting perpendicular to the direction in which the wave propagates.

Imagine a circular ring or hoop in a plane at right angles to the direction of the source. As the waves pass, the hoop is distorted into an ellipse. It contracts in one direction and is stretched in the other. The previously shorter axis then stretches, and the larger one gets squeezed. This pattern of alternate squeezing and stretching, in

A gravitational wave causes a distortion of space that travels at the speed of light. Any body in the path of the wave would feel a tidal gravitational force that acts perpendicular to the wave's direction of propagation. Thus a circular hoop, intercepting the wave face-on, would be distorted into an ellipse. Waves could be detected by monitoring the relative positions of four masses, one in each quadrant (or indeed of any three of these, forming an "L").

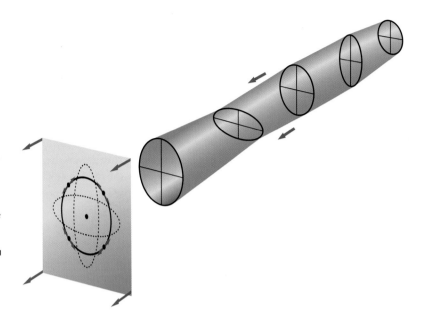

two directions at right angles to each other, is the characteristic signature of gravitational radiation. It is as though the ring of matter is feeling a tidal gravitational force that reverses during every cycle of the wave. However, an ant crawling on the hoop during one cycle would erroneously conclude that the source of the tidal force was in a direction perpendicular to the direction of the true source of the waves.

To detect such a wave requires equipment that can sensitively record tiny changes in distance – and the effects are indeed tiny. When the two neutron stars in the Hulse–Taylor binary pulsar finally merge, the intense pulse of gravitational waves, lasting a few seconds, will cause fractional changes in distance of only one part in 10^{17} as measured at the Earth. The distance between two particles 100 meters apart would change by less than the diameter of an atomic nucleus. But, assuming that we can measure such tiny changes in distance, we still have to wait 300 million years before the system ends its life in a burst of gravitational radiation. There are other similar binaries in our Galaxy, but probably not enough to give more than one coalescence every hundred thousand years. Few experimenters derive satisfaction from building ultrasensitive equipment that detects nothing in an entire lifetime – one event per year is a minimum incentive. To achieve this goal they need equipment sensitive enough to record a coalescence not just in our own Galaxy, but in any of the nearest 100 000 galaxies. The typical such event would be ten thousand times farther away than the binary pulsar: the amplitude of the motions induced by the waves would

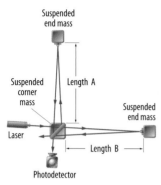

Suspended
end mass

Suspended
corner
mass

Length A

Suspended
end mass

Laser

Length B

Photodetector

In an interferometric gravitational wave detector, a laser beam is split and reflected off mirrors attached to two end masses lying several kilometers away, in perpendicular directions. A gravitational wave causes the two lengths A and B to change: A shortens relative to B, and then B shortens relative to A. There will be a consequent change in the interference pattern between the two reflected beams when they are recombined. The detection of gravitational waves is a daunting experimental challenge because the expected change in path length amounts to only about ten trillionths of the wavelength of light!

then be ten thousand times lower. A gravity wave detector must therefore be able to measure fractional distance changes of one part in 10^{21}!

This is a daunting experimental challenge. Early efforts concentrated on designing "bar detectors" – large metal bars, to which "strain gauges" are attached that can detect the motions if the bar is set vibrating by a passing gravitational wave. There clearly would be spurious "events" caused by earthquakes and other local disturbances, so two or more detectors, at different locations, would be needed to pick out genuine gravitational waves. Although some experimenters are developing bars, it is hard to make them sensitive enough.

A more promising technique uses laser beams to measure the distance between two masses. In the United States, the Laser Interferometer Gravitational–Wave Observatory (LIGO) uses laser interferometers with a 4-kilometer baseline; one is in Washington State, the other in Louisiana. Similar detectors are operating in Europe and Japan. In 2007, LIGO contributed to studies of gamma-ray bursts by failing to detect gravitational waves from a burst that appeared to be located in the nearby Andromeda Galaxy. Short gamma-ray bursts are thought to arise from the mergers of neutron stars or black holes in binaries, which would produce easily detectable gravitational waves at such a small distance. Assuming that the detectors were operating correctly, the non-detection therefore implies that the burst was actually much farther away than Andromeda (and appeared in the same direction on the sky by coincidence), or that the binary merger model is incorrect.

After its next major upgrade – planned for completion around 2013 – LIGO should achieve enough sensitivity to detect the final stages of coalescing neutron star binaries out to 300 million light-years – far enough to have a good chance of detecting one event per year. If there were also a population of binaries consisting of black holes that were, say, 10 times heavier than a neutron star, LIGO could "see" their mergers at even larger distances. LIGO enthusiasts emphasize that other types of as yet unknown sources might be discovered. But there is a fundamental limitation on the kind of "gravitational wave astronomy" that can be done from the ground. Terrestrial detectors are sensitive only to high-frequency waves, those oscillating 10 to 100 times per second. The frequency declines as the mass of the colliding objects rises, so only coalescing bodies with a total mass below a few hundred suns would trigger a response from these detectors. Even if a low-frequency detector were built on the ground, seismic background noise would mask the signal. Thus, LIGO will not be able to detect mergers of very massive black holes, because the waves would vary too slowly. Such mergers

Aerial photograph of the LIGO (Laser Interferometer Gravitational-Wave Observatory) detector near Hanford, in Washington State; the other is near Livingstone, Louisiana. Each arm of the interferometer is 4 kilometers long. It is hoped that an upgraded version of these instruments will detect the coalescence of neutron star binaries, mergers of stellar-mass black holes, and supernova explosions that are sufficiently nonsymmetrical.

are expected among the supermassive holes in the nuclei of galaxies. And, fortunately, their gravitational wave signals are detectable – but only from space.

Supermassive Black Hole Binaries

Our discussion so far may have given the impression that single massive black holes at the centers of galaxies are always "home grown," either introduced as the seed around which the galaxy formed, produced in the collapse of a dense star cluster, or grown by accretion during the first episode of quasar activity. But such black holes can also be acquired all at once by hole-less galaxies, and can even be lost by galaxies. Many, perhaps even most, galaxies have experienced a merger with another galaxy since the first generation of massive holes formed – indeed, merging galaxies are quite commonly observed. Whenever two galaxies merge, the orbits of their constituent stars get mixed up, and the resultant "star pile" resembles an elliptical galaxy. If each of the original galaxies contained a massive black hole, these holes would spiral together into the center of the merged galaxy, forming a supermassive black hole binary.

Kip Thorne, of the California Institute of Technology, was a student of John Wheeler at Princeton. Thorne and his "school" have been responsible for many of the key developments in relativistic astrophysics. For thirty years, he has been the prime promoter of the LIGO project.

The holes gradually approach each other, as they transfer kinetic energy to stars and perhaps suffer drag on gas. At some point they are close enough that each is influenced more by the gravitational pull of the other than by the collective effect of the surrounding stars. The black hole binary continues to shrink, until the two holes are sufficiently close for gravitational radiation to carry energy away faster than the stars or gas can. During these latter stages of binary evolution, the rotation axis of each black hole can precess in the gravitational field of the other. If a pair of jets is produced by one of the holes, the jets should also precess – this precession, or an abrupt change in spin orientation when the black holes finally merge, may explain the "inversion symmetric" radio sources discussed in Chapter 6. (A similar explanation was tried for the jet-producing X-ray binary SS 433, but ran into difficulties. It is more likely that the precession in SS 433 arises mainly from properties of the ordinary star and accretion disk, rather than the compact object.)

Observations have so far identified few convincing candidates for supermassive black hole binaries. In the strongest case to date, two nuclei in the radio galaxy 0402+379 appear to be separated by only 25 light-years. X-ray, optical and radio observations of the galaxy NGC 6240 show a pair of active nuclei 100 times farther apart, but destined to become a close binary in the distant future. The spectacular pair of radio trails in the galaxy cluster Abell 400, once thought to emanate from twin nuclei of the single radio galaxy 3C 75, are now suspected of coming from the nuclei of two separate galaxies. But there are indications that these galaxies are in the early stages of an interaction that could eventually cause them to merge.

A powerful burst of gravitational radiation is emitted during the final coalescence. Since that burst is likely to be focused toward one direction, at the moment of final coalescence the merged black hole may experience a recoil – a strong-field gravitational effect that depends on a lack of symmetry in the system. Numerical simulations suggest that typical recoil speeds may exceed several hundred kilometers per second, which could be enough to propel the merged hole clear out of the galaxy! At the very least, it would eject the black hole from the nucleus. The hole would then gradually regain its central location by losing energy to the stars in the galaxy.

Binary black holes can be ejected from their host galaxies even before the final coalescence occurs. If a third hole drifts in before the binary has merged, it will form a triple system with the existing pair of black holes. "Three-body" systems are notoriously unstable according to Newtonian physics, and a "slingshot" effect can lead to the ejection of *all three* holes, with a "kick" comparable to the orbital velocity of the

The product of a merger between two smaller galaxies, NGC 6240 contains two actively accreting black holes separated by about 3000 light-years. The holes will eventually spiral in to the center of the merged galaxy and form a binary. In this image from the *Chandra X-ray Observatory*, blue indicates the most energetic photons.

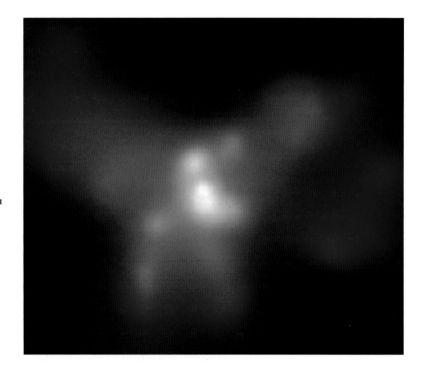

binary, several thousand kilometers per second. The disconcerting possibility exists that a large number of massive black holes could now be hurtling through intergalactic space.

Suppose that two merging galaxies each harbor a supermassive black hole. As the holes spiral together toward coalescence, the energy equivalent to millions of solar masses could be emitted in gravitational waves. How could we detect this colossal gravitational wave energy? Since no force other than gravity is at work, nothing in the physics differentiates a merger between two black holes of, say, 10 solar masses each from a merger between two black holes of a billion solar masses each. One merger is simply a scaled-up version of the other – radii, wavelengths of gravitational radiation, and timescales are all just proportional to the masses. The size of a black hole scales with its mass; so also does the minimum orbital period just before coalescence, which determines the frequency of the waves radiated. When supermassive black holes coalesce, the periods are hours rather than milliseconds, and thus the wave frequencies are low rather than high. As we discussed earlier, waves of such low frequency leave no sign as they pass through ground-based gravitational wave detectors. But they can be detected by a system of spacecraft, such as the *Laser Interferometer Space*

The radio galaxy 3C 75 appears to contain two active nuclei, each of which produces a pair of jets. The projected separation is about 10 000 light-years – small compared to the size of the optical galaxy. Astronomers now believe that the nuclei are actually situated in separate galaxies that are lined up on the sky, but that the galaxies are physically close to each other and may be in the early stages of a merger. The radio image (pink) is superimposed on an X-ray image showing hot gas in the galaxy cluster Abell 400 (blue). Interactions with this gas may lead to the similar flow patterns in both pairs of jets.

Antenna (*LISA*) project being studied by NASA and the European Space Agency.

LISA would consist of three separate spacecraft deployed in a huge equilateral triangle, at separations of 5 million kilometers, all orbiting around the Sun. Small changes in the relative positions of the spacecraft, caused by passing gravitational waves, would be monitored by transmitting laser beams between them. This array could easily detect a merger of two black holes of a million solar masses each, even if it occurred in a very distant galaxy. *LISA*'s sensitivity would be so great that it could follow in detail how the waves gradually strengthen and the orbital motion speeds up as the black holes spiral together.

Long before *LISA* is launched, theorists will have calculated the precise wave patterns expected from coalescing black holes, and how these depend on the masses, spins, and orientations of the two holes. Astrophysicists will be able to compare these calculations with *LISA* data to probe phenomena that are dominated by strong-field

An artist's impression of the *LISA* gravitational-wave detector in space. The array of three spacecraft would go around the Sun in the same orbit as the Earth, but about 20 degrees away. The separations between the spacecraft (about 5 million kilometers) would be monitored by laser beams. This instrument, free of the seismic interference that bedevils terrestrial instruments like LIGO, would be sensitive to gravitational waves of low frequency, such as those that arise from orbits around supermassive black holes.

gravity, and thereby obtain "clean" tests of the crucial predictions of Einstein's theory. The discouraging news for experimenters is that on average a merger between two supermassive black holes takes place less than once per decade over the entire observable Universe.

Fortunately, *LISA* need not sit idle between the rare mergers of quasar-size holes. If there is a large population of intermediate-size black holes, of ten thousand to a million solar masses, then these should merge with greater frequency. Moreover, because of its great sensitivity, *LISA* can detect the fainter signals from objects that might be much more common than supermassive black hole binaries. There are binary stars

Predicted sensitivity of *LISA*, the planned space interferometer, to various gravitational-wave sources. The orange band shows the minimum signals that should be detectable at different wave frequencies (given in Hertz, or cycles per second); the yellowish area to the left is also unobservable due to overlapping signals from millions of close binary stars in our Galaxy. In the light blue area near the bottom of the curve, separately measurable signals from thousands of white dwarf binaries throughout the Galaxy will be seen. The vertical scale is the wave amplitude multiplied by the square root of the number of wave periods collected during a year of observations, or the remaining time before merger of the binaries if that time is shorter. (A weak signal is easier to distinguish from background noise if it can be summed over many periods.) Arrows show the evolution of signal strength and frequency from the spiraling together, and eventual coalescence, of black hole (BH) binaries 10 billion light-years away (at "cosmological" distance). These events are easily detectable. Even the inspiral of a 10 $M_\odot$ hole into a $10^6 M_\odot$ hole is detectable. (The multiple arrows plotted for this case show the range of signals depending on the spin of the holes.)

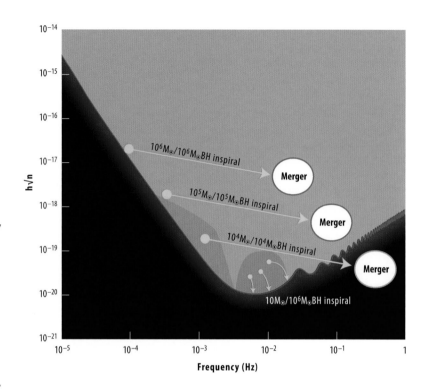

in our Galaxy with orbital periods of about an hour; the nearest of these should be detectable. A more interesting prospect is that *LISA* might detect the waves from low-mass compact stars orbiting supermassive black holes. The waves emitted in this case are, of course, much weaker than those emitted by a supermassive binary – the amplitude is less by the ratio of the star's mass to the hole's mass. But *LISA* may be sensitive enough to detect the almost-periodic signals from such stars.

Is General Relativity the Right Theory?

Einstein's theory dates from 1915, but even fifty years later empirical support for the theory was meager. Two of its predictions – the bending of light near the Sun and the precession of Mercury's orbit – had been tested, but only with about 10 percent accuracy. Several rival theories were still being taken seriously in 1965. But the evidence in support of general relativity is now much more compelling. In our Solar System, there have been amazingly precise measurements of how radio waves are bent near the Sun, of the small extra time delay of radar echoes from planets, and of deviations in the orbits of space probes. These have confirmed that the Sun slightly "curves" the surrounding space exactly as predicted – theory and observation now match to better than one part in 10 000. Moreover, the gradual shrinkage of the binary pulsar's orbit due to gravitational radiation confirms a more subtle and distinctive prediction of Einstein's theory.

We are now more confident that Einstein's theory can be extrapolated into the "strong-gravity" regime relevant to black holes. But we have seen in this chapter that, despite our confidence, there is yet no direct confirmation that general relativity applies under the most extreme conditions – nor that black holes are described exactly by Kerr's equations.

Astrophysicists believe that active galaxies are powered by black holes, just as stars are powered by nuclear fusion. Einstein's theory stands in the same relation to quasars and the centers of galaxies as nuclear physics does to ordinary stars. When astrophysicists model stars, they feed in the relevant nuclear physics as they understand it to be, then calculate the stars' properties and check them against observations. With detailed enough stellar models, however, one could also carry out this process in reverse – using astronomical data to check whether the nuclear physics is right, or even to learn some new physics. This has happened in one case: the dearth of neutrinos detected from the Sun, compared to theoretical predictions, led to the discovery that neutrinos change from one form to another and therefore must have mass. Likewise, those who try to model galactic nuclei (and their small-scale counterparts, the X-ray sources described in Chapter 3) are hoping to probe the fundamental physics of strong-field gravity. They may eventually be able to do so, but "relativistic astrophysics" is at a much earlier stage of development than stellar structure theory, so the goal is still remote.

The renaissance in gravitational research that began in the late 1960s was stimulated partly by the use of new and more powerful mathematical techniques (which led to the proof of the "no hair" theorems), but also by the realization that astrophysical phenomena may actually exist where the effects predicted by the general theory of relativity are large. Theories of gravity can be tested fully only by astronomical observations and by experiments in a cosmic environment, rather than by laboratory work. In the short run, the most useful tests of rival gravitational theories may well come from further high-precision experiments in the Solar System. But in order to study situations where gravity is overwhelmingly strong, one must look farther afield. The highly suggestive evidence already obtained surely justifies the intensive study of black holes using all promising techniques of space- and ground-based astronomy.

11

THROUGH THE HORIZON

Spacetime
singularity

Time

Horizon

Infalling matter
on worldlines

◄── Space ──►

In this "spacetime diagram," the trajectories represent the "worldlines" discussed in the text; worldlines passing each red dot must lie within the attached "light cone." Near a collapsing mass, the light cones are tilted: light can more readily move inward than outward. Material that collapses to form a black hole is crushed in a central singularity, but this singularity is shrouded from view by a horizon. Light emitted just outside the horizon takes a very long time to reach a distant observer. Inside the horizon the light cones are so strongly tilted that even outwardly directed radiation is pulled in toward the singularity.

The black holes we have discussed in this book – mainly the supermassive ones in galactic nuclei, but also the stellar-mass black holes that are the evolutionary endpoint of some stars – are *faits accompli*. We clearly see their manifestations, and with a few diversions our main task has been to describe what they are like. There is nothing mysterious about the origins of the stellar-mass holes – they are the inescapable fates of stars too massive to settle into retirement as white dwarfs or neutron stars. The origins of the hugely massive holes in galactic centers are less certain, but we can be confident that much of their growth, at least, is attended by the spectacular luminosity of quasars or the vast jets of radio galaxies, which are easily visible.

Black holes tend to be thought of in apocalyptic terms that make it difficult to consider them as being "commonplace," at least as much so as stars and galaxies. The 1979 Walt Disney Productions film, *The Black Hole*, typifies popular conceptions in portraying a place of doom, a *terra incognita* at the edge of the Universe. We have tried to dispel such overromanticized notions by concentrating on the concrete ways in which black holes have been sought and found, and the ways in which they affect their environments.

Although we view the evidence for black holes with equanimity, we also recognize that black holes *are* extraordinary features of the Universe. While general relativity gives us a nearly perfect description of their interactions with the outside world, their interiors – where all paths must converge toward a point at which physics as we know it breaks down – fill even jaded astrophysicists with awe. To say that the Universe is sprinkled with black hole singularities – each shrouded by a horizon – is quite a statement indeed.

A useful concept for characterizing the trajectory of a particle is to describe its "worldline" through "hyperspace," that is, its trajectory through space and time. There is a lot of hyperspace in the Universe, and the chance of any two worldlines intersecting by design is small. For example, suppose you were playing interstellar billiards and you wished to hit the eight ball with the cue ball from a distance of 30 000 light-years. It would not be easy. But if the cue ball and the eight ball were both falling inside the same black hole, their worldlines would be *guaranteed* to meet (albeit not under conditions conducive to the continuation of the game). Black holes are places where many worldlines converge and are, figuratively speaking, tied together into a knot from which they can never escape. Black holes are therefore catchment basins where matter and energy are "taken out of circulation" and can no longer interact with the rest of the Universe *except* gravitationally.

On a less poetic level, we can ask whether the gravitational collapse of matter to form black holes has broader implications than "merely" the creation of quasars, radio jets, and X-ray binaries. This final chapter considers black holes from the viewpoints of both cosmology and quantum physics – the realms of the very large and the exceedingly small. Did black holes form in other ways than through the collapse of stars or the coalescence of material in the centers of galaxies, as described in Chapters 2 and 4? How much might they contribute to the dark matter, which makes up more than 80 percent of a typical galaxy's mass? And will it ever be possible to create a black hole in a particle accelerator on Earth?

Pregalactic Black Holes?

Edwin Hubble, in the 1930s, discovered that our Universe is expanding – because of this expansion, the light from quasars and distant galaxies is redshifted. In 1965, Arno Penzias and Robert Wilson made an equally momentous discovery while scanning the sky with a radio antenna at the Bell Telephone Labs at Holmdel, New Jersey. (This was the second major serendipitous discovery from that laboratory – Carl Jansky's pioneering studies in radio astronomy, discussed in Chapter 6, were carried out there as well.) Penzias and Wilson found that the Universe is pervaded by dilute "heat radiation" having no identifiable source. Cosmologists quickly interpreted this radiation as a relic of the so-called "hot big bang," when the very young Universe was filled with intense radiation and squeezed to immense densities and temperatures.

The "primordial" radiation can never disappear – it fills the Universe and has nowhere else to go. But it does become cooled and diluted during the expansion, so that now, 14 billion years after the big bang, its temperature is -270 degrees Celsius (less than 3 degrees above the "absolute zero" temperature). Over the forty-plus years since its discovery, the "cosmic background radiation" has been probed with increasing precision, most notably by the instruments on board NASA's *Cosmic Background Explorer (COBE)* and *Wilkinson Microwave Anisotropy Probe (WMAP)* satellites, as well as a number of ground-based experiments. These measurements show that the radiation has exactly the form expected if it is indeed an afterglow of the big bang.

During the first second of its expansion, the Universe would have been hotter than 10 billion degrees – far hotter even than the center of a supernova explosion. How did this primordial fireball evolve into our present Universe of galaxies? If the Universe had started off absolutely uniform, there would be no stars, no galaxies, and, worst of all, no cosmologists. Its present-day complexities

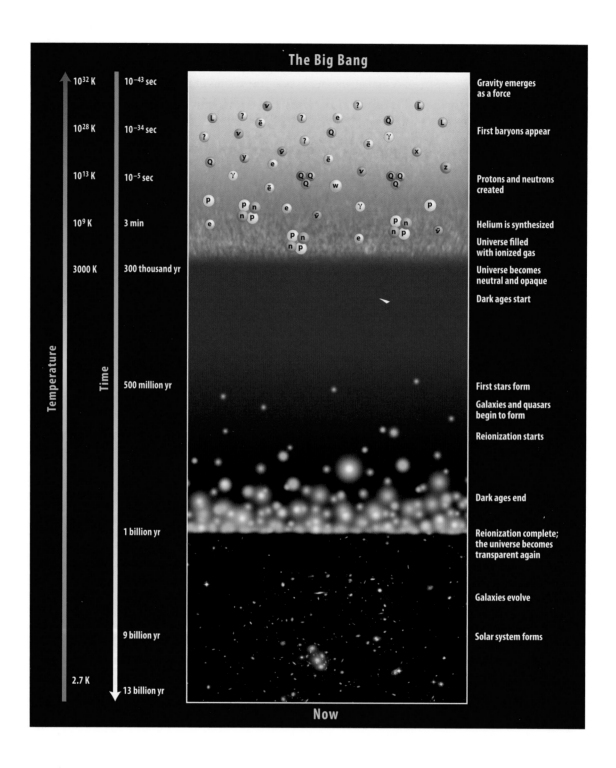

couldn't have emerged unless the Universe had been slightly irregular right from the start. Galaxies and other structures could then have evolved from patches that started off slightly denser than average, or were expanding slightly more slowly than average – as we discussed in Chapter 4. The expansion of these patches would have lagged further and further behind their surroundings, and eventually their excess gravity would have halted their expansion, allowing them to condense.

The larger the initial "overdensity," the sooner gravity overcomes the expansion. The early Universe would have been "roughest" on small scales; that is, on these scales the density fluctuations would have been most marked. In other words, if you had averaged the density over larger and larger volumes, the early Universe would have appeared increasingly smooth. This is not a surprising characteristic of the fluctuations – you would get a similar effect if you measured a coastline on a very fine scale, then "averaged" it over larger distances in order to draw a large-scale map. This graininess has an important consequence for the development of structure in the Universe. It means that, were gravity the only relevant force, small scales would condense first. A cloud of matter with the mass of a star, for example, would condense before a cloud with the mass of a galaxy. The resultant objects would then cluster on progressively larger scales, stars gathering to form galaxies, galaxies gathering to form clusters of galaxies, and so on. This "hierarchical clustering" scenario formed the basis for our discussion of galaxy formation in Chapter 4.

This is close to but not exactly what happened. Gravity doesn't act alone – both the gas and the cosmic background radiation exert pressures as well. When the Universe was less than 100 000 years old, the energy of the background radiation so dominated the mass-energy of matter that condensation was suppressed. By the time matter started to dominate, the smallest structures had been smeared out, while on larger scales the early Universe was smoother, so that it took longer for the excess gravity to reverse the initial expansion. We therefore have good reason to believe that the first cosmic structures to form had an intermediate mass equivalent to about a million suns.

The *COBE* and *WMAP* satellites were able to observe this condensation process in its early stages, by comparing the temperature of the background radiation on different parts of the sky to a precision of one part in 100 000. These measurements revealed slight non-uniformities. Since the radiation is a direct messenger from early pregalactic epochs, these temperature fluctuations – the much-hyped "ripples in space" – directly confirm that the early Universe indeed possessed the kind of irregularities from which cosmic structures could have evolved.

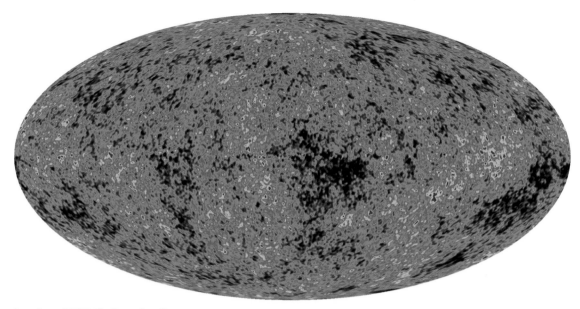

Data from *WMAP*. The "snapshots" taken by *COBE* and *WMAP* show the Universe as it was several hundred thousand years after the big bang. A thick screen of electrons, which scatters the cosmic background radiation much like fog scatters a car's headlight beams, prevents us from seeing any further back in time. The strong radiation field at earlier times also would have washed out the smallest irregularities and prevented any mildly dense patches of larger size from condensing.

It is hard to compute what happens inside one of these pregalactic structures after it starts to collapse. The gas could fragment into a large number of separate stars, some of which subsequently coalesce; or much of it could collapse as a single entity, forming a supermassive star that quickly evolves into a black hole. In either case, a substantial fraction of the gas in the Universe could in principle have collapsed into black holes, at an early epoch before galaxies formed. These black holes could have had masses of anywhere from a few hundred to a million suns, and might have begun forming when the Universe was a few tens of millions of years old.

Along with other forms of dark matter, these black holes would have clustered, creating progressively larger gravitationally bound structures. After a billion years, they would have built up to the scale of galaxies. If the remaining gas fell into these systems and turned into stars, we would then end up with galaxies embedded in dark "haloes" replete with black holes. As we discussed in Chapter 4, each galaxy is indeed embedded in an extensive halo of dark matter, but indications are that most of this dark matter is "non-baryonic," that is, not composed from familiar forms of matter. Still, it is possible that up to 10 to 20 percent of the dark matter could consist of black holes much heavier than stars – each perhaps as heavy as a million suns.

Whenever one of these black holes passed through the central parts of the galaxy, where the stars are packed relatively close together, it would transfer some of its energy to the (much lighter) stars. This energy transfer would act like a friction that slows down the hole's motion. Black holes that spend too much time in this star-laden environment may suffer so much "drag" that they spiral into the center. The first two to reach the center would form a binary; they may then spiral together and coalesce, as a consequence of their release of gravitational radiation. When a third hole falls in, the process repeats, and so on. In this way, a single very massive hole may form. But the outcome is very different if a third hole drifts in toward the center before the binary has coalesced. The resultant three-body system is unstable and disrupts very quickly: a binary shoots off one way, and the third hole recoils in the opposite direction, leaving nothing behind in the galactic center at all. If black hole binaries were always disrupted by a third body before they had been driven to coalescence, then a galaxy could have zero, one, or two holes in the center at any given time, but these would never accumulate into a heavier one by merging.

If black holes exist before galaxies do, and can settle toward the centers of galaxies forming among them, they could serve as seeds for the formation of AGN. Once present, a central black hole could grow through repeated mergers or the accretion of gas during the quasar epoch. Mergers and accretion represent two alternative routes to the buildup of large central black hole masses.

Black holes of thousands (rather than millions) of solar masses could also exist, as remnants of an early generation of very massive stars. As we discussed in Chapter 2, black holes may form from ordinary massive stars, although it is hard to calculate which types of supernovae give rise to neutron stars and which leave behind black holes. However, any "superstar" weighing more than 300 solar masses would certainly become a black hole – indeed, it would end its life by collapsing without any spectacular explosion. We know that stars as heavy as this are not found now. However, there could have been a bigger proportion of such objects among the first generation of stars.

Many astrophysicists, notably Bernard Carr of London University, have explored how many of these "very massive objects" there might be. It turns out that they could be important contributors to the dark matter. In contrast, we can assert quite confidently that black holes in the ordinary stellar-mass range (from a few solar masses up to about 200) can't be major contributors to the dark matter. They have been ruled out because the formation of each lower-mass hole would have been accompanied by a supernova explosion. Every time a supernova occurs, several solar masses of "processed" material – containing carbon, oxygen, and iron – is ejected into space. If these holes contributed even 1 percent of the dark matter in our Galaxy's halo, their precursors would have produced more of these chemical elements than we observe. So, at most, only a tiny fraction of our halo could consist of black holes that are remnants of the kinds of star that exploded as supernovae. The same can be said for black holes more massive than about a million solar masses – they would have revealed themselves through their gravitational effects on other objects in the Galaxy. On the other hand, there is no such constraint on holes between about 300 and a million solar masses: they swallow, rather than expel, the material processed by their precursor "superstars," and are not massive enough to strongly influence Galactic structure.

Thus there could be enough black holes, with any mass between 300 and a million suns, to account for much of the baryonic component of dark matter in the haloes of galaxies. How could we detect such a population? One of the most promising techniques makes use of gravitational lensing, already discussed in Chapter 4. A black hole's gravitational field would deflect light rays passing close to it, and a distant quasar could appear magnified and distorted if one of these holes lay in an intervening galaxy along the line of sight; the lensing could also produce multiple images of the distant source. For black holes at the upper end of the allowed mass range, around a million solar masses, the images would be separated by a milliarcsecond – a tiny angle, but one that radio astronomers can resolve using very-long-baseline interferometry.

Black holes of more modest mass can also disclose themselves by lensing distant objects; but it is easiest to detect these holes if they are in our own Galactic halo. If a black hole in the Milky Way's halo moves across the line of sight to a background star, its gravity focuses the light so that the star appears brighter. The magnification rises to a peak (when the alignment is closest) and then declines again; so the background star brightens and fades in a predictable way. Unfortunately, such an event shouldn't happen very often – alignment must be very close to get the effect. Suppose we look at a background star, and ask what the chance is that a black hole lies along the line of sight to it, sufficiently well lined up to cause significant magnification. Even if there were enough black holes to make up all the dark matter in our Galaxy, the chance is only about one in 10 million.

To stand a chance of detecting the effect, one must monitor millions of background stars, measuring their brightnesses on a regular basis. Until the late 1990s, merely handling the reams of data generated by this approach would have been too daunting. Since then, automated observing programs have become routine. These programs pick up thousands of variable stars of many types. Many astronomers make a living out of studying pulsating stars, flare stars, and binaries; but for this purpose the intrinsic "variables" are a nuisance. The challenge is to pick out the rare instances of variability that manifest the characteristic symmetrical rise and fall of a lensing event, and which are also "achromatic," in the sense that the amplitude of variability is the same in blue and red light. Numerous lensing events have been detected – mainly involving objects with masses less than that of the Sun – but the frequency of these events is not enough to account for more than about a tenth of the halo mass. Black holes are harder to find, because the "size" of the lens, and the time it takes for the brightness of a lensed object to rise and fall, goes up as the square root of the lens's mass. So any events caused by the kinds of black hole that might contribute significantly to the dark matter in our Galactic halo would be at least 20 to 30 times slower than the events detected to date. It is therefore not surprising that no convincing examples have yet been found.

Miniholes and the Hawking Effect

We have seen that there are circumstances under which black holes of a star's mass or higher could form immediately before, during and after the era of galaxy formation – an era that started a few tens of millions of years after the big bang. But what about the possibility that black holes formed much earlier? Observations of the cosmic background radiation offer a "snapshot" of the Universe as it was several hundred thousand years after the big bang. A fog of electrons – which scatters the background radiation much as earth-bound fog scatters a car's headlight beams – prevents us from seeing further into the past. Thus, we cannot explore directly the processes that occurred during these "primordial" times (although gravitational waves emitted during this period would penetrate the fog and might someday be detectable). Nevertheless, we can say something about how and when such black holes *might* have formed – if, indeed, they *did* form.

Under primordial conditions, the tiny fluctuations responsible for the formation of stars and galaxies would not be adequate to trigger black hole formation. Something more dramatic would have had to happen to cause collapse. Physicists struggling to understand the behavior of matter and energy under the extreme conditions of the big bang have proposed a variety of effects that might lead certain regions to recollapse quickly while the rest of the Universe continued to expand. The size of a primordial black hole would depend mainly on when it formed. Black holes forming as "late" as one second after the big bang would weigh about a hundred thousand solar masses, while those collapsing earlier would have been proportionately smaller: solar-mass black holes could have formed about

10^{-5} seconds after the big bang. A primordial era of black hole formation could not have lasted much longer than a second; if it did it would have interfered with the production of helium during the first few minutes.

A primordial black hole weighing a few solar masses would be identical to a black hole formed via the recent demise of a massive star. But any black hole weighing much less than a star would have to be primordial. These could have formed, if at all, only in the ultradense initial instants of the big bang – the era when temperatures and densities were far higher than can ever be attained in a laboratory, even inside a particle accelerator. These "miniholes" are especially interesting to physicists because they may yield fundamental insights into how gravity links to the other forces of nature.

Finding a "unified theory" that joins all types of forces is perhaps the most ambitious aim of physicists. Gravity is well described by Einstein's general theory of relativity – undoubtedly one of the great achievements of twentieth-century physics. But Einstein's theory, like Newton's, is a "classical" theory. It doesn't incorporate the concepts of quantum theory, which revolutionized and deepened our understanding of atoms and forms the basis for modern descriptions of electromagnetism and the forces within atomic nuclei. A key tenet of quantum theory is Werner Heisenberg's famous "uncertainty principle." The speed, or the energy, of a particle cannot be pinned down instantaneously: the more precisely we wish to know these quantities, the longer the measurement takes.

The lack of a fully "quantized" version of Einstein's theory hasn't impeded progress in astronomy. Gravity is so weak that it can be ignored on the scales of atoms and molecules, where quantum effects are important. Conversely, when gravity is important – in planets, stars, galaxies – the scales are so large that we can ignore quantum effects. Thus, in ordinary circumstances there is no real overlap between the domains in which these two great pillars of physics, general relativity and quantum theory, are relevant. But developing a theory of "quantum gravity" is conceptually of great importance for physicists, and it is of immediate relevance for astrophysicists and cosmologists. Such a theory is needed before physicists can develop a fully "unified" theory of all the forces. In the absence of such a theory we will never have a consistent picture of the very first instants of the big bang (when the density would have been so high that quantum effects were crucial for the entire Universe), or of what happens near the "singularity" deep inside a black hole, where conditions become just as extreme.

There is no accepted theory of quantum gravity – indeed, there is not even a consensus among specialists about the most promising path toward it. An important first step, however, has been taken. Relativists have developed an approach that takes account of quantum effects on the particles and radiation in space, but still treats gravity, and space itself, by Einstein's theory. This "hybrid" approach – quantizing the particles and radiation in a "classical" gravitational field – led Stephen Hawking to a remarkable new insight that linked Einstein's relativity with quantum concepts; as a bonus, he discovered unexpectedly deep connections between gravity and thermodynamics. Freeman Dyson rated Hawking's conceptual breakthrough as "one of the great unifying ideas in physics."

Hawking's results apply in principle to any black hole, but their consequences would be significant only for very tiny ones – naturally, given the link to quantum theory. To introduce the "Hawking effect," let us consider the properties of holes with different masses. A black hole's radius is proportional to its mass, but according to Einstein's theory there is no qualitative difference between holes of low and high mass – the former are just miniature versions of the latter. Some other object, or some other physical process apart from gravity, must come into the picture to set a scale. For instance, when a star orbiting a massive black hole feels a tidal force, the outcome depends sensitively on the size of the black hole relative to the star: black holes exceeding a billion solar masses are so big that they could swallow a star whole; a star falling toward a somewhat smaller hole would be torn apart first, creating a more conspicuous display.

A black hole weighing as much as the Sun would be 6 kilometers across; one with the mass of the Earth would have a diameter of only 9 millimeters. What would happen if there were black holes the size of atoms? Such holes would still be enormously heavier than atoms. A hole as big as an atomic nucleus would weigh as much as a small asteroid. To make such a hole, about 10^{36} protons would have to be packed into the dimensions normally occupied by one. This particular large number should come as no surprise. We note in Chapter 1 (and also in the Appendix) that the gravitational force attracting two protons is only 10^{-36} of the electric force repulsing them. As many as 10^{36} protons must therefore be packed together before gravity can compete with the vastly stronger forces that hold protons apart.

We might expect quantum effects to be important around such a small hole. But how do these effects arise, and what is their connection to gravity? The key lies in the nature of the tidal force near the black hole. Supermassive holes in the centers of galaxies pull more strongly on the near side of a passing star than on its far side; this differential (or "tidal") force tears apart any star that comes too close to the hole. The tidal forces are fiercer around miniature, atomic-scale black holes – so much so that even individual electrons and protons are perturbed by them.

Empty space, what physicists call "the vacuum," is seething with activity on the microscopic quantum scale. When high-energy particles crash together, new particles are created, along with their antiparticles. Such collisions, observed routinely in particle accelerators, may create a shower of electrons and positrons, for instance. These electrons and positrons, and indeed all other possible kinds of particles and antiparticles, can be thought of as latent everywhere in space, as "virtual" particles. Normally, virtual particles can exist only very transiently, because the Heisenberg uncertainty relation allows them to "borrow" their rest-mass energy for only a very short time. A virtual proton–antiproton pair, for example, can exist for only about 10^{-24} seconds before disappearing back into the vacuum. However, these pairs can be made "real" – that is, long-lived – if they can acquire enough energy during the time of their brief appearance. For instance, an electron and positron can be permanently "created" if an electric field pulls them apart strongly enough to supply their rest-mass energy before the Heisenberg relation demands that it must be paid back.

The strong gravity near a black hole can create particles in a similar way, except that the force that pulls the particles apart is the tidal gravitational force – a miniature version of the effect that tears a star apart around a supermassive hole. By chance, one virtual particle will appear closer to the black hole than the other. If the tide is strong enough, it can pull these particles apart so violently that they acquire their entire rest-mass energy mc^2 within the brief span of time for which the uncertainty relation allows this energy to be "borrowed." The particles then need not annihilate to repay the energy debt: they instead transform into a real particle and a real antiparticle. But there is a catch – they can't both escape from the hole. One escapes; the other is trapped inside the hole with, in effect, *negative energy*. What do we mean by the expression "negative energy"? When a black hole swallows a particle that comes in from far outside, it grows in mass. Such particles always have positive energy. But absorbing a particle with negative energy actually diminishes the hole's mass: the other particle, carrying positive energy away from the hole, may escape the hole's clutches. Hawking's remarkable inference was that black holes actually emit something. They glow, rather than being completely "black," and in doing so they gradually lose mass. Thus, a black hole isolated in space will actually "evaporate"!

Hawking's mechanism permits the creation only of particles that are as large or larger than the hole itself. According to quantum mechanics, the effective size of a particle is inversely proportional to its mass. Only very tiny holes, therefore, can emit electrons and positrons. Holes of any mass, however, can emit photons: the only requirement is that the wavelength of the light must be larger than the size

Stephen Hawking's greatest contribution was his realization, in 1974, that black holes are not completely black but radiate with a well-defined temperature. This finding has no significance for the holes that have actually been discovered, but it would be astrophysically important if "miniholes" exist. Hawking's discovery revealed deep conceptual links among gravity, quantum theory, and thermodynamics. Earlier, he had been responsible, with Roger Penrose, for showing that singularities form inside black holes, even when the collapse happens in a nonsymmetrical way. His more recent researches have dealt mainly with the consequences of quantum theory for the entire Universe.

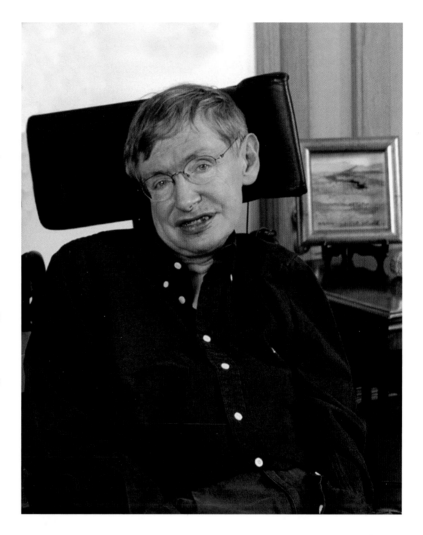

of the hole's horizon. Depending on the hole's mass, the dominant radiation could be X-rays, visible light, or microwaves. Just as a hot ingot of steel glows red, orange, or blue-white, according to its temperature, so Hawking reasoned that black holes must glow with a well-defined temperature, proportional to the gravitational force just outside the hole.

Before Hawking derived his result, several people had applied the concepts of thermodynamics to black holes. It was known that the surface area of the horizon around a black hole could increase if, for instance, things fell into the hole and it gained mass. But that same surface area could never decrease; if two holes merged, the horizon around the resultant hole would have a bigger area than the two original holes added together. To some physicists this sounded a lot like the Second

The worldlines of virtual particles close to a small black hole. Particle pairs, consisting of every possible particle and its antiparticle, are constantly appearing and disappearing at each point in space. Under normal conditions, Heisenberg's uncertainty principle allows these particles to exist only very briefly. Close to a black hole's horizon, however, tides can pull them apart with such force that they acquire enough energy to last indefinitely. When this happens, one particle of the pair is always swallowed by the hole, having lost so much energy that it actually decreases the hole's mass; the other escapes to infinity. Through this analysis, Stephen Hawking discovered in 1974 that black holes would emit heat radiation and could ultimately "evaporate."

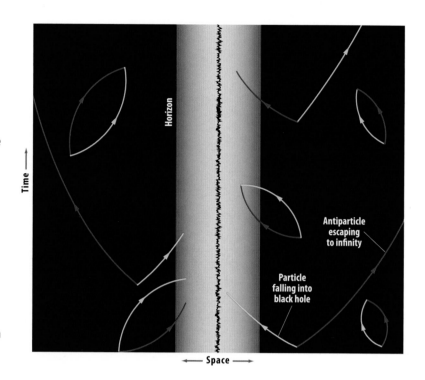

Law of Thermodynamics, which states that the "entropy" of a system – a measure of its disorder – can never decrease. They speculated that, somehow, the area of a black hole's horizon was analogous to entropy. They could hardly have imagined how profound this analogy would prove to be.

When we say that the disorder, or entropy, of a system is increasing, we mean that information is being lost about how the system was set up. That lost information is irretrievable. When equal volumes of hot and cold gas are mixed together, for example, the increase in entropy is related to the amount of information that is lost about the microscopic properties of the gas. Before mixing, all the slow atoms might be confined to one half of the volume, and all the fast ones to the other half. Afterward, they are inextricably mixed, and we can no longer make that distinction. According to general relativity, the formation of a black hole entails a similar loss of information. When a black hole forms or grows, all trace disappears of what fell into it. We can think of not just the objects themselves disappearing into the black hole, but also all information about those objects. A 10-solar-mass black hole created by the collapse of a burnt-out star's iron core is *identical* to a black hole that has grown to the same mass by accreting hydrogen or swallowing pineapples. The "entropy" of a black hole can thus be characterized by the amount of information lost in making it; amazingly, the area of its horizon is a precise measure of this loss.

The Israeli physicist Jacob Bekenstein (then one of John Wheeler's students at Princeton) pushed this analogy further. Still reasoning from the principles of thermodynamics, he pointed out that if the area of a black hole resembled entropy, then the strength of gravity on its surface must resemble temperature. But he was uneasy about whether this insight could be more than just a metaphor, because black holes were then presumed to be truly "black," in the sense of absorbing radiation but emitting none. Bekenstein's insight was vindicated when Hawking showed that the surface isn't completely black but radiates with a well-defined temperature that is indeed proportional to the strength of gravity at its surface.

Hawking's ideas on "black hole evaporation," proposed in 1974, are now generally accepted. They were, however, so contrary to earlier intuitions that they proved hard to assimilate at the time. When Hawking presented his work for the first time, at a conference in Oxford, the session chairman was overtly disparaging, and soon afterward published a paper "refuting" the idea. It took more than a year for some leading experts – people such as Yakov Zel'dovich and his colleagues in Moscow – to become convinced of Hawking's claims.

Evaporating Black Holes

The theory of black hole evaporation, sometimes called the "quantum radiance" of black holes, has been "battle tested" in the sense that it has been rederived in many different ways by different people. But that is no substitute for actually observing the predicted radiation, as perhaps we could if very small black holes actually existed. Whether or not the predicted effects can be observed, the Hawking effect ranks as a pinnacle in our understanding of gravity. But is there a chance of actually observing the "quantum radiance" of a black hole?

The predicted temperature is proportional to the gravitational force just outside the hole. This force depends on the mass of the hole divided by the square of its radius. The radius of a hole scales with its mass M, so the "Hawking temperature" will be large for small holes and small for large ones. The black holes that form when massive stars die would be only a millionth of a degree above absolute zero. For these (and for the even bigger black holes in galactic nuclei), Hawking's process is utterly negligible: big holes radiate energy far more slowly than they soak it up from the primordial radiation that pervades even intergalactic space. Only if the Universe were to continue to expand for 10^{66} years (enormously longer than the current age of 10^{10} years) would stellar-mass holes evaporate.

But the temperature of a "minihole" the size of a proton is a billion degrees, high enough for it to radiate X-rays and gamma rays. This radiation would be detectable if such an object came anywhere near our Solar System. As a minihole radiates, it shrinks, and its radiation gets hotter and still more intense, until the minihole finally disappears in a flash of gamma rays. (It is controversial among theorists whether it disappears completely, or leaves some kind of particle as a relic.)

The fabric of space is punctured by black holes, each marking the death of a star or the outcome of some runaway catastrophe in the center of a galaxy. But no astronomical processes operating today could create a black hole smaller than 2 or 3 solar masses. Any star of lower mass, even if it had exhausted all nuclear sources, could survive indefinitely, quite stably, as a neutron star or white dwarf. It could transform into a black hole only if a huge external pressure squeezed or imploded it to even higher densities than those of a neutron star. Anything smaller than a star – say, a planet or asteroid – would need to be squeezed to a higher density still.

John Wheeler, in the 1960s, envisaged how miniholes could be created by the artificial "implosion" of a suitable mass to colossal densities. Wheeler's idea has lately generated renewed interest, though still in the spirit of "science fiction." Physicists have speculated that space inside a black hole, rather than being crushed indefinitely toward a "singularity," could instead sprout into an entire new universe, expanding almost from nothing into a space of its own, quite distinct from and inaccessible to our Universe. According to the popular "inflationary" cosmology, our own Universe may have actually formed in a similar way. Alan Guth of the Massachusetts Institute of Technology, the leading early proponent of "inflation," has written semi-facetiously about "creating a universe in the laboratory."

If we were to find a minihole, however, we need not interpret it as the artifact of a supercivilization: it could be a relic of the early Universe. In the first microsecond of its expansion, our Universe was far denser than a neutron star. The tremendous pressures could have created much smaller black holes than are produced astrophysically at current epochs. Whether this would have actually happened depends on how chaotic or irregular conditions were.

It is not sufficient that the pressure be high: pressure *differences* between one place and another are necessary to drive an implosion. The best current guess about the early Universe suggests that the pressure variations would have been too smooth. Minihole formation would have required the early Universe to have been very much choppier on small scales than we would estimate by extrapolating from the larger scales at which galaxies form.

The gravitational effect of one of these miniholes would be undiscernible unless it came within our Solar System. This would be an unlikely event even if there were enough miniholes to make up all the dark matter in the halo. The Hawking radiation itself offers the best hope of detecting miniholes. Lower-mass holes are hotter and radiate more brightly, and we can conclusively exclude the possibility that the dark matter is made up of miniholes with masses below 10^{18} grams (corresponding to a size of 10^{-10} centimeters) simply because they wouldn't be "dark" enough: their combined radiation would make the sky brighter in X-rays and gamma rays than it is observed to be. But miniholes of this tiny size could still of course exist in smaller numbers, without violating observational limits.

Black holes that formed in the big bang with masses far below 10^{15} grams would not have survived to the present epoch: they would have "evaporated" long ago. Those that started off with masses of 5×10^{14} grams would just now be in their final stages of evaporation. The lifetime goes roughly as the cube of the hole's mass; those that started off with twice the mass of the holes now disappearing would be only an eighth of the way through their lives. Each of these miniholes – with a mass of 10^{15} grams, but smaller than a single atom – could radiate 10 gigawatts of gamma rays for 10 times the lifetime of the Solar System. Even though we can rule out there being enough of these holes to contribute to the dark matter, it would be extraordinarily interesting to find even one of them. These entities are a truly remarkable creation of pure thought – a rigorous deduction from the two best-established theories in twentieth-century physics: quantum electrodynamics, combined with general relativity.

Miniholes are possible "fossils" of the ultra-early Universe, "missing links" that might illuminate the quest for a theory unifying particle physics and gravitation theory. And they are eminently detectable, if enough of them exist. In their death throes, they would emit gamma rays intense enough to be detected many light-years away. (However, the gamma rays from evaporating holes don't have the right properties to account for the observed gamma-ray bursts described in Chapter 7.) The details of the final explosions are sensitive to uncertainties in the details of particle physics. According to some theories, the final explosion may eject an intense fireball of electrons and positrons. There would then be an even more sensitive way of seeking evaporating miniholes. Sweeping up the ambient magnetic field as it plowed into the surrounding matter, the fireball would emit a flash of radio waves. Radio telescopes could detect such an event, caused by an entity smaller than an atomic nucleus,

The Large Hadron Collider, near Geneva, Switzerland, is the world's most powerful particle accelerator. Some speculative theories involving extra dimensions of space predict that the LHC might create microscopic black holes. According to quantum theory, such a black hole would evaporate almost instantly; even if it managed to survive, its tiny mass would render it harmless.

even if it happened in the Andromeda Galaxy – two million light-years away.

Black Holes in the Laboratory?

While the quantum mechanical size of a particle is inversely proportional to its mass, the Schwarzschild radius increases with mass. Thus, according to quantum mechanics, a black hole with a small enough mass should spill over its own event horizon and cease to be a black hole. This limiting mass, roughly 10^{-5} grams (about 10^{19} times the mass of a hydrogen atom), is called the "Planck mass" after the German physicist Max Planck. It is generally considered to be the minimum mass that a black hole can have.

Is it possible to create a Planck-mass black hole in a particle accelerator? Unfortunately (or fortunately, depending on your level of caution), the answer is no. The Planck mass, small as it seems, corresponds to an energy about a thousand trillion times higher than the maximum energy attainable by the world's most powerful accelerator, the Large Hadron Collider located near Geneva, Switzerland. Even the

The unusual characteristics of gravity, compared to the other forces of Nature, may imply that there are more than three spatial dimensions. In order to avoid easy detection, these extra dimensions would have to be "compactified," or curled up so that they are too small to see. In the simple example of compactification shown here, a two-dimensional plane (top) can be rolled into a tube, the radius of which can be shrunk until it resembles a line – a one-dimensional object (second row). Likewise a torus (third row) can be compactified to resemble a point (bottom).

most energetic cosmic rays – particles that enter the Earth's atmosphere from space and can be studied through their interactions with the atmosphere, also fall far short.

But all may not be lost. An abiding puzzle about gravity is its extreme weakness compared to the other forces of nature. These forces – electromagnetism, and the strong and weak nuclear forces – have been "unified" by the recognition that they increasingly resemble one another, and become comparable in strength, as the distance between interacting particles becomes extremely small and their collision energy becomes correspondingly large. If gravity followed the same kind of unification, it would mean that Newton's gravitational constant, G, must become larger as particles are pushed closer together. In other words, the gravitational force between two particles would go up faster than the inverse square of their separation, once the separation is sufficiently small. This would expand the size of the Schwarzschild radius (which is proportional to G) and imply that the minimum mass of a black hole is much smaller than the Planck mass.

One way to explain the unification of gravity with the other three forces is via the existence of extra spatial dimensions, beyond the familiar three. As early as 1918, the German physicist Theodor Kaluza suggested that gravity could be unified with electro-magnetism if there were a fourth dimension of space (or a fifth, if one includes time); Oskar Klein, of Sweden, later proposed that the extra dimension could be "hidden in plain sight" – virtually undetectable – if it were curled up, or "compactified," into a very tiny circle. Modern extra-dimensional approaches to unification require as many as 10 spatial dimensions, 7 of which would have to be compactified!

If the force of gravity spreads through extra dimensions, while the other forces are confined to, say, the familiar three, then this dilution could explain why gravity appears so weak. The full force of gravity would not be seen until one explored scales smaller than the sizes of the compactified dimensions. The distances (and energies) that would have to be probed would depend on the number of extra dimensions accessible to gravity, as well as their sizes. For example, the Large Hadron Collider could create a black hole if gravity were spread through two extra dimensions, provided that these dimensions had sizes as large as a fraction of a millimeter. (If gravity spreads through a larger number of dimensions, then a smaller compactification length scale would suffice.) There is no compelling theoretical reason to suppose that such "large extra dimensions" must exist, but the idea of creating a black hole in the laboratory – and proving the existence of extra dimensions – is so breathtaking that even improbable scenarios are worth exploring.

Lisa Randall, of Harvard University, is a leading proponent of the possibility of extra dimensions. In work with Raman Sundrum, she suggested that our entire Universe could be embedded in a space of higher dimension. This idea gives a natural explanation for why gravity is such a weak force on the atomic scale.

A black hole created in the laboratory would behave in a dramatically different fashion from other kinds of particles created in accelerators, and should be easy to detect. It would decay, via the Hawking effect, into a "fireball" consisting of all possible kinds of elementary particles. The possibility that a hole of such tiny mass would survive more than an infinitesimal fraction of a second is extremely remote, but even if it did, there is no danger that it would "destroy the Earth," as alarmists have suggested. If such a black hole escaped from the laboratory, it could pass through the Earth unnoticed. Even if it became trapped by the Earth's gravitational field, it would hardly interact with anything, oscillating through the Earth's core while occasionally swallowing a proton. In any case, if quantum black holes are capable of being created in an accelerator, then cosmic rays must have been creating them in the atmosphere since the Earth was formed – apparently with no ill effects.

Looking Forward

The history of structure in the Universe is the story of how gravity gradually overwhelms all other forces. First, the original cosmic expansion must be defeated: because some parts of the Universe start off slightly denser than average, or expand slightly slower than average, they condense into structures held together by their own gravity. If the early Universe had been very irregular on small scales, this condensation process could have happened very early, giving rise to primordial black holes.

Systems condensing later would become protogalaxies, in which gas is gradually transformed into successive generations of stars. In individual stars, gravity is balanced by pressure and nuclear energy; low-mass stars, even when their fuel runs out, can survive as white dwarfs or neutron stars. But the cinders of very massive stars collapse all the way to black holes. These holes are hard to detect, except when they are orbited by a companion star that provides a supply of fuel – they are then revealed by their emission of X-rays. In the centers of galaxies, runaway processes lead to the formation of supermassive holes that sometimes manifest themselves by generating the tremendous luminosity of quasars, or by squirting out the jets of strong radio sources.

The amount of matter locked up in black holes will continue to grow into the long-range cosmic future: more and more black holes will form; existing holes will grow by capturing gas, radiation, and, in the case of supermassive holes, even entire stars. Popular books sometimes overdramatize this process by asserting that everything in our Galaxy will eventually be engulfed by the central black hole. This is not realistic. Our Sun, and most of the material in the Galaxy, has so much angular momentum that it will never approach close enough to the center to be endangered. The Sun is far more likely to be kicked out of the Galaxy by close encounters with other stars.

What about the fate of our entire Universe? All the evidence suggests that it will expand indefinitely. If so, it remains to be determined whether the expansion will continue to accelerate, or will level off and eventually slow down (but never quite stop). In an ever-expanding Universe, stellar-mass black holes would continue to form, until all the gas was tied up in dead stellar remnants. The fate of galaxies and clusters of galaxies, however, would depend on the future rate of acceleration. If acceleration turns out to be a shortlived phase, then the galaxies would not evolve in isolation. Rather, they would merge into larger systems. After a trillion years, the typical "galaxy" would be an amorphous system of stellar remnants (including stellar-mass black holes) as massive as a supercluster of galaxies is today. In a Universe with perpetual acceleration, however, the future will be lonelier. The Local Group – our Galaxy, Andromeda, and the Group's smaller members – will merge into a single amorphous system. But all other galaxies and clusters will accelerate away, and eventually fade from view.

Nowadays, Hawking's evaporation process is important only for microscopic black holes, but in a perpetually expanding Universe even the largest holes would be affected eventually. Given enough time, they would release, in the form of radiation, all the matter and energy they had ever swallowed. Stellar-mass black holes would take 10^{66} years to evaporate. Supermassive holes, the remnants of long-dead quasars, would take even longer – more than 10^{90} years for the largest so far discovered in galactic nuclei. But the fate of black holes depends, too, on the future of cosmic acceleration. If the acceleration fades out, they might never finally disappear – indeed they could become ever larger. Long before they could be eroded by evaporation, black holes in the nuclei of merging galaxies would spiral inward and coalesce into still larger ones at the centers of the merged systems. This process could happen repeatedly as the scale of structure grew. The time it takes a hole to evaporate depends on the cube of its mass. Black holes could grow, by repeated coalescence, always remaining big enough to survive however long the Universe continued to expand.

And what of the future of black hole research?

Astronomers will continue to flesh out the demographics of the supermassive black holes that form in galactic centers. Knowing the distribution of their spins as well as masses will help us to understand how they formed and grew; within a decade or two we may be able to detect the gravitational waves emitted during their mergers. The existence of remote quasars tells us that some of these black holes have existed for more than 90 percent of cosmic history, and many pieces of evidence suggest that they regulated the galaxy formation process through a feedback mechanism that remains to be understood. New generations of telescopes may reveal these black holes in their early stages of growth, while improvements in computing should provide a better understanding of how they affect their environments.

Demographic surveys, as well as improved computer models, will also lead to a better understanding of how stellar-mass black holes form. Which massive stars collapse to black holes, and which disperse entirely or leave behind neutron star remnants? Gamma-ray bursts are prime suspects as sites for black hole formation: future gravitational wave detectors may prove or disprove this hypothesis. Observable out to redshifts approaching 100, bursts are potentially important probes of early cosmic structure – even if we don't understand them fully.

When provided with fuel (from, for instance, a close binary companion star), both stellar-mass and supermassive black holes manifest intense emission over a broad range of wavelengths; sometimes, they expel jets. We will continue to seek a better understanding of the physical processes going on in both accretion disks and jets. The challenge for astrophysicists is not just to understand in detail the complex ways in which these populations of black holes manifest themselves, but also to find new ways of testing whether the holes actually have the properties that Einstein's theory predicts.

Gravitational lensing may reveal whether a third black hole population, intermediate in mass beween the other two, exists in the haloes of galaxies. These black holes might be remnants of an early generation of very massive stars, and could have formed at any stage when the Universe was between a million and a billion years old – after it has cooled down sufficiently for objects to condense, but before galaxies are themselves fully formed. Whether such "intermediate-mass" black holes exist depends on uncertain details of how the first generation of stars formed, and on the nature of the irregularities in the early Universe. There is, however, nothing unnatural about scenarios that would allow much of the gas in the early Universe to condense into black holes. An important population of black holes – perhaps even the overwhelming fraction of those that now exist – could therefore predate the first galaxy.

A fourth population – miniholes – would be a missing link between the cosmos and the microworld; the quest for these extraordinary objects is worthwhile even if the chance of finding them is

"The response of astrophysicists to a fashionable new idea."

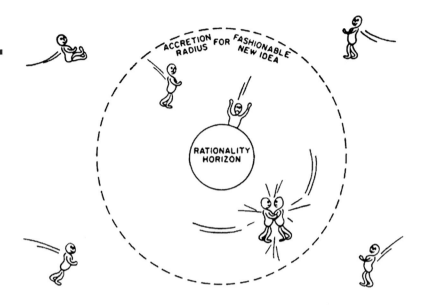

small. We would ourselves, however, bet long odds against their existence. They seem unlikely ever to have been created, because their formation requires rather contrived initial conditions, which are hard to reconcile with other constraints on what the early Universe could have been like. They are the least likely category of black holes to exist.

Finally, the remote possibility of creating black holes in the laboratory focuses attention on the fundamental links between black holes and the frontiers of physics. Are there more dimensions of space than meet the eye? Does a black hole really retain no vestige of its provenance, other than mass and spin – or does it subtly encode the process by which it formed and reveal all when it evaporates? Are the singularities of black holes truly dead ends, or can they become gateways into new, expanding universes?

Few would have asked such questions thirty years ago, when it was still considered slightly risqué to assume that accreting black holes power quasars. In a celebrated lecture in 1977, the astrophysicist Richard McCray lampooned the ascendance of the black-hole "bandwagon" in quasar research by describing "the response of astrophysicists to a fashionable new idea." Using an accretion metaphor, he described how the astrophysicists are first attracted to the idea when they wander into its sphere of influence. Many fly through without getting trapped. Occasionally, however, two astrophysicists collide, generating a great deal of heat – but not always much light. These hapless souls can no longer leave the sphere of influence of the fashionable new idea, but are fated to slip through the "rationality horizon," within which the idea ceases to be an interesting hypothesis and becomes an article of faith.

Today, an astrophysicist who *doubted* that black holes lie at the heart of the quasar phenomenon, and many other phenomena besides, would be regarded as venturing dangerously close to the "rationality horizon." The importance of black holes in cosmic phenomena is no longer a subject for speculation; that threshold has moved to more esoteric realms. Black holes may or may not turn out to hold the key to the unification of forces, or the origin and fate of our Universe. But it is a safe bet that the "fatal attraction" of gravity will continue to draw us toward deeper insights about the nature of the cosmos.

Appendix: Gravity and Cosmic Dimensions

The force of gravity is described by an inverse square law, just like the electrical attraction between positive and negative charges. We can get a "feel" for the relative strength of gravity by considering its effect in a single hydrogen molecule, containing two protons. The gravitational attraction between two protons is only 10^{-36} as strong as the electrical attraction, which is of course why atomic physicists can safely ignore it. But every gravitating body has the same "sign" of gravitational charge: there is no near-cancellation of positive and negative, as there is for the electrical forces within any macroscopic body. Gravity can therefore become significant when huge numbers of atoms are packed together.

We can measure gravity's importance by working out the "gravitational binding energy" per hydrogen atom. This quantity, discussed in Chapter 1 in the cases of globular clusters and planets in the Solar System, measures the amount of energy needed to move a particle away from a mass to a large distance. The gravitational binding energy of a particle orbiting a mass M at a distance r depends on M/r. When N atoms are packed together in a sphere (at a given density), the mass M will be proportional to N (and thus M/r to N/r), and the radius r will be proportional to $N^{1/3}$. The gravitational binding energy per atom (proportional to N/r) therefore goes as $N/N^{1/3}$, equal to $N^{2/3}$. Gravity starts off with a "handicap" compared to the electrical forces that hold atoms together. But as we consider larger and larger aggregations of atoms, it gradually "catches up": specifically, it gains by a factor of 100 for each thousandfold rise in N, and becomes dominant when N is as large as $10^{36 \times 3/2} = 10^{54}$. This simple argument encapsulates the basic physical reason why stars are as massive as they are: a cold body containing more than 10^{54} hydrogen atoms (or protons) would be compressed, because gravity is too strong to be balanced by the same interatomic forces as in ordinary solids. A cold body with N larger than 10^{57} is so completely crushed that it forms a black hole. (This is the famous "Chandrasekhar limit.") A hot body of this mass can become a hydrogen-burning star.

These and other "numerological" relationships can be instructively displayed on a diagram such as the figure on page 276. In this diagram, mass is plotted vertically and radius horizontally. The plots are given on a log scale: an increase by a factor of 10 in mass or radius is the same fixed length everywhere on each axis. Masses are given as multiples of the proton mass and lengths as multiples of the proton radius. The other distinctive feature of the diagram is that it spans an enormous range: the vertical (mass) axis goes from a single proton up to a mass that is nearly 80 powers of ten larger.

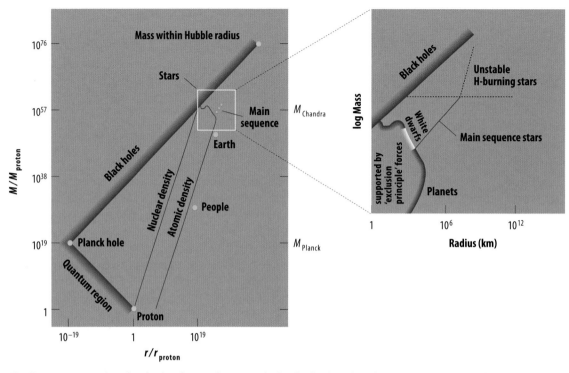

The diagram summarizes the physics of stars, planets, and other bodies in a plot of mass against radius. Ordinary solid lumps lie on the line marked "atomic density": the mass, plotted in units of the proton mass, scales as the cube of the radius. That line would eventually cross the "black hole" line at a mass of about 100 million suns, except that it is curtailed before it can do so. The reason is that any object containing more than about 10^{54} atoms (about the mass of Jupiter) would be crushed by gravity to a higher density than an ordinary solid. The smaller "inset" diagram shows in more detail the limited range of scales in which stars can exist. Along the thin red diagonal line, the central temperatures (related to M/r by the virial theorem) are high enough for hydrogen fusion to occur. Main sequence stars lie along this line. Lower-mass stars need to become denser than higher-mass stars to achieve a given central temperature. No object below about 0.07 solar masses can burn hydrogen because the degeneracy pressure resulting from the exclusion principle would prevent it getting hot and dense enough. Stars of more than about 100 solar masses are supported mainly by radiation pressure, which makes them vulnerable to instabilities.

The relation between the mass and radius of a black hole is shown in the diagram by a line with slope 1. Nothing can exist to the left of this line; anything very far to the right of it has such a large radius, in relation to its mass, that gravity is unimportant. Note that a black hole the size of a proton weighs as much as 10^{38} protons. Miniholes have to be so heavy simply because, for a single proton, gravity is weaker than the microscopic forces by just the inverse of that number. (Chapter 11 explains why holes with this particular mass are especially interesting.)

We can also plot on the same graph the mass–radius relation for objects of fixed density. These lines have slope 3 (mass being proportional to the cube of radius). Two such lines are shown. The left-hand one corresponds to the density of an atomic nucleus. It meets the "black hole" line at a mass of 10^{57} protons: neutron stars, confined by gravity to dimensions not much larger than their Schwarzschild radii, are indeed around this mass.

The right-hand line corresponds to objects of "ordinary" density, where the atoms, rather than their central nuclei, are "touching." It is parallel to the nuclear density line, but displaced to the right by about $100\,000$ – this is the factor by which an atom is larger than its nucleus. This straight line extends only up to about 10^{54} protons, which is roughly the size of Jupiter. In bodies of larger mass, gravity starts to overwhelm the ordinary atomic forces that hold solids together, crushing such bodies to higher densities. The behavior becomes especially interesting as the mass approaches 10^{57} protons. This part of the diagram is shown magnified at the right, and the legend explains the physical processes further. We can readily see, for instance, why stable stars (gravitationally bound fusion reactors) can exist only for a range of masses around 10^{57} protons.

The diagram extends up as far as the mass of the observable Universe (i.e., the mass within the Hubble radius). This mass is of course very poorly known, but it is about 10^{76} protons – the square of the basic 10^{38}. This is actually not a coincidence. It can be shown that the lifetime of a star is about 10^{38} times longer than the time light would take to cross an atom. So by the time the Universe is old enough for stars to have evolved in it, its "size" (the Hubble radius, or the distance light has traveled since the big bang) must be 10^{38} times larger than an atom. The Universe had to be old by the time we came on the scene; it therefore (if it is homogeneous) has to be large as well.

Also plotted in the diagram is a line of slope -1, which represents the uncertainty in the location of a mass (or, in a sense, its minimum size) set by quantum theory. This meets the black hole line for a mass of 10^{-5} grams (10^{19} protons) and a radius of 10^{-33} centimeters (10^{-19} proton radius). Known respectively as the Planck mass and the Planck length, these are the dimensions at which quantum and gravitational effects overlap. The Planck scales lie many powers of ten away from the "ordinary" scales confronted in the laboratory, or even in astronomy. We can also define a basic timescale, the Planck time, by dividing the Planck length by the speed of light. A big bang cosmological model could not be validly extrapolated back to the Planck time, whose value is 5×10^{-44} seconds, until it is understood how quantum effects operate over the entire Universe.

The important message of the figure is that the characteristic masses and radii all involve simple powers of a single large number, whose vastness is a consequence of the weakness of gravity in the microworld of atoms. It is because gravity is so weak that structures dominated by its effects have to be so massive. If gravity were not quite so weak, the basic shape of the diagram would be unchanged, but the horizontal and vertical scales would be compressed: they wouldn't stretch over so many powers of 10. In a hypothetical universe where gravity was stronger, cosmic structures could be neither so large nor so long lived. But, if the gravitational force were zero, no cosmic structures could exist at all. Paradoxically, the weaker gravity is (provided it isn't exactly zero), the grander are its consequences.

Further Reading

Note: Numbers in parenthesis denote relevant chapters.

Audouze, Jean, and Guy Israel, eds. *The Cambridge Atlas of Astronomy*. 3rd ed. New York: Cambridge University Press, 1994.

Bartusiak, Marcia. *Einstein's Unfinished Symphony: Listening to the Sounds of Space-Time*. Joseph Henry Press, 2000. (10)

Begelman, Mitchell. *Turn Right at Orion: Travels Through the Cosmos*. Cambridge, MA: Perseus, 2000. (2, 3, 4, 5, 6, 8)

Friedman, Herbert. *The Astronomer's Universe*. New York: W. W. Norton, 1990. (3)

Greene, Brian. *The Elegant Universe: Superstrings, Hidden Dimensions, and the Quest for the Ultimate Theory*. New York: W. W. Norton, 1999. (11)

Hawking, Stephen W. *A Brief History of Time*. New York: Bantam Books, 1988. (1, 11)

Hawking, S. W., and W. Israel, eds. *Three Hundred Years of Gravitation*. New York: Cambridge University Press, 1987. (1, 10)

Kaler, James B. *Stars*. Scientific American Library. New York: W. H. Freeman, 1992. (2)

Katz, Jonathan I. *The Biggest Bangs: The Mystery of Gamma-Ray Bursts, the Most Violent Explosions in the Universe*. New York: Oxford University Press, 2002. (7)

Luminet, Jean-Pierre. *Black Holes*. New York: Cambridge University Press, 1992. (1, 10)

Melia, Fulvio. *The Edge of Infinity: Supermassive Black Holes in the Universe*. New York: Cambridge University Press, 2003. (4, 5, 8)

Novikov, Igor. *Black Holes and the Universe*. New York: Cambridge University Press, 1990. (1, 3)

Randall, Lisa. *Warped Passages: Unraveling the Mysteries of the Universe's Hidden Dimensions*. New York: Harper-Collins, 2005. (11)

Rees, Martin. *Before the Beginning: Our Universe and Others*. (Helix Books) Addison-Wesley, 1997. (4, 5, 11)

Rees, Martin. *Perspectives in Astrophysical Cosmology*. New York: Cambridge University Press, 1995. (4, 5, 11)

Schilling, Govert. *Flash!: The Hunt for the Biggest Explosions in the Universe*. Naomi Greenberg-Slovin, trans. Cambridge, UK: Cambridge University Press, 2002. (7)

Silk, Joseph. *A Short History of the Universe*. Scientific American Library. W. H. Freeman, 1997. (4, 5, 11)

Sparke, L. S., and J. S. Gallagher, III. *Galaxies in the Universe: An Introduction*. 2nd ed. New York: Cambridge University Press, 2007 (4. More advanced.)

Taylor, Edwin F., John Archibald Wheeler, and Edmund Bertschinger. *Exploring Black Holes: Introduction to General Relativity.* 2nd ed. New York: Addison-Wesley, 2008. (1, 4, 6, 10. More advanced.)

Thorne, Kip S. *Black Holes and Time Warps: Einstein's Outrageous Legacy.* New York: W. W. Norton, 1994. (1, 3, 10)

Verschuur, Gerrit L. *The Invisible Universe Revealed: The Story of Radio Astronomy.* New York: Springer-Verlag, 2006. (6)

Wheeler, John A. *A Journey into Gravity and Spacetime.* Scientific American Library. New York: W. H. Freeman, 1990.

Wheeler, J. Craig. *Cosmic Catastrophes: Exploding Stars, Black Holes, and Mapping the Universe.* 2nd ed. New York: Cambridge University Press, 2007. (2, 3)

Will, Clifford M. *Was Einstein Right? Putting General Relativity to the Test.* 2nd ed. New York: Basic Books, 1993. (10)

Figure Credits

Chapter 1

page

4   Hebrew University of Jerusalem Albert Einstein Archives, courtesy AIP Emilio Segrè Visual Archives. Contact www.albert-einstein.org about publishing this image.

7   Photograph by Robert Bein, courtesy AIP Emilio Segrè Visual Archives.

12   Image supplied courtesy of Roy Kerr.

14   Reproduced by kind permission from the Syndics of Cambridge University Library, shelf mark Rel.f.1.b.48.

16   ESO PR Photo 20/06, The Globular Cluster 47 Tuc (FORS/VLT).

20   Cartoon by Sidney Harris, © ScienceCartoonPlus.com, Reproduced with permission from S. Harris.

Chapter 2

28   NASA, ESA, M. Robberto (Space Telescope Science Institute/ESA) and the Hubble Space Telescope Orion Treasury Project Team.

30   AIP Emilio Segrè Visual Archives, gift of Subrahmanyan Chandrasekhar.

34   NASA, ESA, J. Hester and A. Loll (Arizona State University).

36   © Anglo-Australian Observatory, photo courtesy of David Malin Images (DMI).

37   Image courtesy of NRAO/AUI and N. Bartel, M. Bietenholz, M. Rupen, *et al.*

40   Images by Leonhard Scheck, H.-Thomas Janka, Ewald Müller (Max Planck Institute for Astrophysics). Published in Stan Woosley & H.-Thomas Janka: "The physics of core-collapse supernovae", *Nature Physics* 1 (2005), 147–154. Reproduced with permission from H.-Thomas Janka and *Nature Physics*.

43   Churchill Archives Centre, The Pulsar Discovery Records of Professor A. Hewish, HWSH.

46   NASA/HST/CXC/ASU/J. Hester *et al.*

47   Adapted from Jacob Shaham, "The oldest pulsar in the Universe," *Scientific American,* 256 (1987), 50.

49   Adapted from Alexander Heger *et al.,* "How massive single stars end their life," *The Astrophysical Journal,* 591 (2003), 288.

Chapter 3

Chapter 4

96 Simulation: John Wise & Tom Abel (KIPAC/Stanford). Visual-ization: Ralf Kähler, John Wise & Tom Abel (KIPAC/Stanford); Reproduced with permission from Tom Abel.

97 Image supplied courtesy of Prof. Lars Henquist. Reproduced with permission from the AAS.

98 Left, J. Trauger, JPL and NASA; right, Hubble Space Telescope WFPC 2 Team and NASA.

102 (a) Image reproduced with permission from Axel Mellinger and EKTECH Köhler GmbH (b) © NASA COBE (DIRBE, DMR, FIRAS).

103 Credit: NRAO and F. Yusef-Zadeh.

103 NASA/CXC/UCLA/MIT/M.Muno *et al.*

Chapter 5

106 AIP Emilio Segrè Visual Archives, Shapley Collection.

107 NASA and The Hubble Heritage Team (STScI/AURA).

108 NASA, ESA, CXC, and JPL-Caltech.

110 Reproduced with permission of Palomar Observatory and California Institute of Technology.

111 R. Perley, C. Carilli & J. Dreher; Image courtesy of NRAO/AUI/NSF.

112 Reproduced with permission of California Institute of Technology.

118 AIP Emilio Segrè Visual Archives, E. E. Salpeter Collection.

119 AIP Emilio Segrè Visual Archives.

120 Image courtesy of Institute of Astronomy, University of Cambridge.

122 Inset: TRACE/Stanford-Lockheed Institute for Space Research.

124 Space Telescope Science Institute, NASA.

Chapter 6

129 Courtesy of D. Jansky and W. T. Sullivan.

130 © Cavendish Astrophysics Group, Cavendish Laboratory.

132 Image courtesy of NRAO/AUI.

135 (a) Digital Sky Survey/U.K. Schmidt Image/STScI; (b) X-ray (NASA/CXC/M. Karovska *et al.*); Radio 21-cm image (NRAO/ AUI/NSF/J. Van Gorkom/Schminovich *et al.*), Radio continuum image (NRAO/AUI/NSF/J. Condon *et al.*); Optical (Digitized Sky Survey UK Schmidt Image/STScI).

136 First published in Gizani, N. A. B.; Garrett, M. A.; Leahy, J. P. "The Multiwavelength Study of Two Unique Radio Galaxies," *Journal of Astrophysics and Astronomy*, **23**, 2002, p. 89; reproduced with per-mission from N. Gizani and J. P. Leahy.

137 Adapted from Bruce Margon, "The bizarre spectrum of SS 433," *Scientific American*, **243** (1980), 54.

139 R682/0004 Artist's impression of double star system SS 433, David Parker/Science Photo Library.

139 Blundell & Bowler, NRAO/AUI/NSF.

140 Gloria Dubner *et al.*, VLA, NRAO/AUI/NSF.

140 C. P. O'Dea & F. N. Owen, NRAO/AUI/NSF.

143 Image courtesy of NRAO/AUI/NSF and Neal Miller, Jansky Fellow of the NRAO.

144 Image courtesy of NRAO/AUI/NSF, M. J. Hardcastle *et al.*

145 Image courtesy of NRAO/AUI/NSF and STScI (inset), D. Merritt and R. D. Ekers.

147 Image courtesy of NRAO/AUI/NSF and F. Owen, J. Biretta & J. Eilek; highest resolution image by C. Ly, R.C. Walker, and J.M. Wrobel.

149 Image courtesy of NRAO/AUI/NSF; Ann Wehrle, *et al.* Image by Glenn Piner.

151 Photograph by Anglo-Australian Observatory.

152 Adapted from Roger D. Blandford, Mitchell C. Begelman, and Martin J. Rees, "Cosmic jets," *Scientific American*, **246** (1982), 124.

153 Image courtesy of NRAO/AUI/NSF, R. A. Laing & A. H. Bridle.

157 Jeff Hester, Arizona State University, and NASA.

159 Adapted from Richard H. Price and Kip S. Thorne, "The membrane paradigm for black holes," *Scientific American*, **258** (1988), 69.

160 Image courtesy of Roger Blandford and Kavli Institute for Particle Astrophysics and Cosmology, Stanford University.

160 S. Hirose, J. H. Krolik, J.-P. De Villiers, and J. F. Hawley, "Magnetically driven accretion flows in the Kerr metric. II. Structure of the magnetic field," *The Astrophysical Journal*, **606** (2004), 1083. Reproduced by permission of the AAS.

162 Reprinted by permission from Macmillan Publishers Ltd: *Nature* **436**, "A dark jet dominates the power output of the stellar black hole Cygnus X-1", Elena Gallo, Rob Fender, Christian Kaiser *et al.*, © 2005. Reproduced with permission of *Nature* and Dr. Elena Gallo.

Chapter 7

164 Adapted from Ray Klebesadel *et al.* 1973, *The Astrophysical Journal*, **182**, L85.

165 Adapted from Peter Mészáros, The Pennsylvania State University.

166 © CGRO Science Support Center, NASA.

167 Photograph courtesy of Princeton University Office of Communications.

168   BeppoSAX Team, ASI, ESA.

169   Images reproduced with permission of Andrew Fruchter.

170   Adapted from Jens Hjorth *et al.* 2003, *Nature*, **423**, 847.

172   Adapted from William Paciesas *et al.* 1999, *The Astrophysical Journal Supplement Series*, **122**, 465.

173   Adapted from NASA's Goddard Space Flight Center.

174   Image by Maximilian Ruffert (University of Edinburgh) and Hans-Thomas Janka (Max Planck Institute for Astrophysics, Garching), reproduced with authors' permission.

175   Images supplied courtesy of Davide Lazzati.

Chapter 8

181   Image reproduced courtesy of Dr. Thomas Ott, Infrared and Sub-millimeter Astronomy Group at MPE.

182   Image reproduced courtesy of Dr. Reinhard Genzel.

183   Image by Rebecca Haycox/UCSB, reproduced with permission of Rebecca Haycox and Dr. Andrea Ghez.

184   Adapted from A. Ghez *et al.*, "High proper-motion stars in the vicinity of Sagittarius A*: Evidence for a supermassive black hole at the center of our galaxy," *The Astrophysical Journal*, **509** (1998), 678.

185   © Adam R. Contos, 2002.

186   These images were created by Prof. Andrea Ghez and her research team at UCLA and are from data sets obtained with the W. M. Keck Telescopes.

186   A. Ghez *et al.*, Keck/UCLA Galactic Center Group.

188   Adapted from M. Miyoshi *et al.*, "Evidence for a massive black hole from high rotation velocities in a sub-parsec region of NGC 4258," *Nature*, **373** (1995), 127.

188   Image by Joh Kagaya (Hoshi No Techou), for Makoto Inoue. Image courtesy of Lincoln Greenhill, Harvard-Smithsonian Center for Astrophysics, reproduced with permission of Makoto Inoue.

189   Adapted from M. Miyoshi *et al.*, "Evidence for a massive black hole from high rotation velocities in a sub-parsec region of NGC 4258," *Nature*, **373** (1995), 127.

194   H. Ford *et al.*, STScI/NASA.

195   T. R. Lauer *et al.*, NOAO/NASA.

196   Adapted from Gary Bower and Richard Green, NOAO/STIS Instrument Definition Team/NASA.

197   Adapted from S. Tremaine *et al.*, "The slope of the black hole mass versus velocity dispersion correlation," *The Astrophysical Journal*, **574** (2002), 740.

199 Reproduced with permission of Jonathan McKinney and Charles Gammie.

200 Adapted from C. R. Evans and C. S. Kochanek, "The tidal disruption of a star by a massive black hole," *The Astrophysical Journal,* 346 (1989), L31.

Chapter 9

205 Reprinted from *New Astronomy Reviews* 50, 9–10, Xiaohui Fan, "Evolution of high-redshift quasars", p.7, (c) 2006, with permission from Elsevier.

206 Data from D. P. Schneider *et al.,* "PC 1247+3406: An optically selected quasar with a redshift of 4.897," *The Astronomical Journal,* 102 (1991), 837.

206 Data from G. Richards *et al.,* "The Sloan Digital Sky Survey Quasar Survey: Quasar luminosity function from Data Release 3," *The Astronomical Journal,* 131 (2006), 2766.

207 Data from G. Richards *et al.,* "The Sloan Digital Sky Survey Quasar Survey: Quasar luminosity function from Data Release 3," *The Astronomical Journal,* 131 (2006), 2766.

208 Data from G. Richards *et al.,* "The Sloan Digital Sky Survey Quasar Survey: Quasar luminosity function from Data Release 3," *The Astronomical Journal,* 131 (2006), 2766.

210 Adapted from B. M. Peterson *et al.,* "Steps toward determination of the size and structure of the broad-line region in active galactic nuclei. III. Further observations of NGC 5548 at optical wavelengths," *The Astrophysical Journal,* 392 (1992), 470.

212 Adapted from D. Croton *et al.,* "The many lives of active galactic nuclei: Cooling flows, black holes and the luminosities and colours of galaxies," *Monthly Notices of the Royal Astronomical Society,* 365 (2006), 11.

213 Adapted from G. Chartas *et al.,* "XMM-Newton Reveals the Quasar Outflow in PG 1115+080," *The Astrophysical Journal,* 595 (2003), 85.

215 Image by Steven Allen (Stanford) and Harald Ebeling (Hawaii), reproduced with permission of Steve Allen.

216 Adapted from J. Peterson *et al.,* "High-resolution X-ray spectroscopic constraints on cooling-flow models for clusters of galaxies," *The Astrophysical Journal,* 590 (2003), 207.

216 Adapted from S. Allen *et al.,* "The X-ray virial relations for relaxed lensing clusters observed with Chandra," *Monthly Notices of the Royal Astronomical Society,* 328 (2001), 37.

217 Credit: NASA/IoA/A. Fabian et al.
219 Credit: NASA/UMD/A. Wilson et al. (both images).
220 Credit: NASA/IoA/A. Fabian *et al.*
220 (a) Illustration: NASA/NASA/CXC/M. Weiss; (b) Credit: NASA/ CXC/W. Forman *et al.*

Chapter 10

226 Adapted from Jean-Pierre Luminet, *Black Holes* (New York: Cambridge University Press, 1992), 143.
227 NASA/GSFC/W. Cash.
227 Christopher A. Perez and Robert V. Wagoner, Stanford University.
228 Figure courtesy of Christopher S. Reynolds, based on data published in A. Fabian et al., "A long hard look at MCG-6-30-15 with XMM-Newton," Monthly Notices of the Royal Astronomical Society, 335 (2002), L1.
229 Adapted from R. A. Remillard and J. E. McClintock, "X-ray properties of black-hole binaries," Annual Review of Astronomy & Astrophysics, 44 (2006), 49.
230 Antony Howarth/ Science Photo Library.
231 Adapted from Kip S. Thorne, Black Holes and Time Warps: Einstein's Outrageous Legacy (New York: W. W. Norton, 1994), 292.
233 Adapted from Kevin Rauch, Canadian Institute for Theoretical Astrophysics (CITA), University of Toronto.
236 Image supplied by Robin Stebbins, NASA Goddard Space Flight Center.
238 Data from J. M. Weisberg and J. H. Taylor, in Binary Radio Pulsars ASP Conference Series, Vol. 328, 2005.
239 Image by C. Henze, NASA.
241 Adapted from R. Ruthen, "Catching the wave," illustrated by George Retseck, *Scientific American,* **266** (1992), 90.
242 Adapted from Kip S. Thorne, *Black Holes and Time Warps: Einstein's Outrageous Legacy* (New York: W. W. Norton, 1994), 383.
243 Image supplied courtesy of LIGO Laboratory.
245 NASA/CXC/MPE/S. Komossa et al.
246 X-ray: NASA/CXC/AIfA/D. Hudson & T. Reiprich *et al.*; Radio: NRAO/VLA/NRL.
247 Image supplied by Robin Stebbins, NASA Goddard Space Flight Center.

248 Adapted from Greg Kuebler, based on data supplied by Peter Bender (JILA, University of Colorado), Robin Stebbins, and Joan Centrella (NASA Goddard Space Flight Center).

Chapter 11

254 Adapted from S. G. Djorgovski and Digital Media Center, Caltech.
256 NASA / WMAP Science Team.
263 © Philip Waterson, LBIPP, LRPS; Image supplied courtesy of Prof. Stephen Hawking.
268 © CERN.
270 Image by Tsar Fedorsky, supplied courtesy of Dr. Lisa Randall.
273 Richard McCray, University of Colorado.

Index

Index

Index

Index

Index

Index

Index

mass and host galaxy (bulge) mass, 196–197, 209, 211, 214
mass and quasar accretion, 210
mass and relativistic orbits, 232
possibility of intergalactic, 245
possible ejection of binary, 244
regulation of cluster growth, 215
regulation of galaxy growth, 212
tidal disruption by, 192, 199–201
X-ray spectrometry of gas near, 226
see also quiescent black holes
supermassive stars, 119, 159, 256
Supernova 1987A, 35–36, 151
Supernova 1993J, 37
Supernova 1998bw, 169, 171
Supernova 2003dh, 170
supernova remnants, 42, 48–49
supernovae
computer simulation, 41
core collapse, 33–36
evolution of galactic nuclei, 99
frequency, compared with gamma-ray bursts, 171
long duration gamma-ray bursts and, 171–172
massive stars and, 31–41
Type Ia, 64
types of, 34
X-ray binary evolution involving, 62
Suzaku satellite, 59
Swift satellite, 172–173, 176–178
symmetry
and black hole formation, 12
of double radio sources, 109, 131, 148
inversion symmetry, 143, 145, 244
mirror-symmetric jets, 142–144
see also asymmetry
symposia. *see* Texas
synchrotron radiation
Crab Nebula, 42, 45
Galactic center, 102–103
GRB and, 177
radio sources, 102–103, 109, 117, 134, 217

T

Taylor, Don, 42
Taylor, Joseph, 237
technology leading AGN theory, 124–125, 128
telescopes
balloon based, 194
see also ground-based telescopes; resolution; satellites; space-based observations

temperature
BH gravity and, 265–266
blackbody radiation and, 39
gravitational collapse and, 18
see also cool cluster cores
Texas Symposia on Relativistic Astrophysics
First symposium, 1963, 52
Ninth symposium, 1978, 137
Seventh symposium, 1974, 164
thermodynamics and gravity, 17, 261, 263, 265
thermonuclear reactions. *see* fusion reactions
Thorne, Kip, 244
three body systems, 244, 257
3C 31, 153
3C 66B, 142, 144, 151
3C 75, 244, 246
3C 273
apparent superluminal expansion, 149
luminosity of, 112
position of, 110
redshift of, 80–81, 112
see also quasars
3C 279, 149
tidal capture binaries, 97
tidal disruption, 192, 199–201, 231
tidal forces and quantum effects, 262
time dilation by extreme gravity, 7
timing argument, 79
Tonry, John, 195
Tremaine, Scott, 76, 195
21cm spectral line, 85–86

U

Uhuru X-ray satellite, 55–56, 58–59
ultraviolet wavelengths, 63, 106, 121, 205, 208–209
uncertainty principle, 260, 262
unification/unified theory
AGN phenomena and line of sight, 121–123
the Hawking effect and, 260, 267
radio galaxies and line of sight, 150–152
singularities and, 11
unipolar inductors, 158
universe
possible creation from BH, 266
primordial condition of, 259
scale of, 72, 277–278
total matter content, 86, 277
see also expansion

Index